Teaching BASIC ENGINEERING MECHANICS

for optimum student involvement

Gregory Pastoll

Copyright and origination

The contents of this book are entirely the work of the author, who retains the copyright in them.

All the illustrations are the work of the author:

Some of the photographs were taken by former students or colleagues, whose names were not recorded at the time.

This work was first published by Gregory Pastoll in 2024

Cover artwork by Gregory Pastoll

Cover design and typeset by Book Covers Australia

bookcoversaustralia.com

ISBN 978-0-6452688-2-9 paperback

ISBN 978-0-6484665-8-1 ebook

Preface

What instructors do in front of a class is far less important than what they get their students to do. I am convinced of this, after more than 30 years of teaching, and observing teaching in a variety of contexts.

For over fourteen years, I taught the subject of basic engineering mechanics, to first- and second-semester students, in a polytechnic/university of technology environment. This was done in two stints: one of five years, and later, one of nine. In between those two periods, I spent another fourteen years as a consultant on teaching methods for the academic teaching staff of the University of Cape Town.

In my capacity as a consultant, I had plenty of opportunity to observe in action and to provide feedback to keen university teachers in a wide range of disciplines.

Back then, the prevailing method of transmitting information to students was by lecturing. Although some of those who consulted me were very good at lecturing, it gradually became clear to me that lecturing was not a method that resulted in active student participation. If people really wanted to improve their teaching, I would advise them to lecture less, and to make use of other activities that would get their students to be more involved, and consequently more motivated.

In promoting this trend, I was inspired by the successes achieved by John Cowan, Emeritus Professor of Learning Development at The Open University, when he taught civil engineering at Heriot-Watt University in Scotland in the seventies.

In my second spell of teaching engineering mechanics, I was able to put into practice a lot of what I had learned as a consultant. I developed a number of 'learning activities' that *did* give my students that kind of experience.

In each successive teaching term I attempted to improve my course. So, what you find in this book is the culmination of a considerable amount of reflective action.

Contents

Preface iii

Introduction vii

Chapters

1. Running effective tutorials 1
2. Exercises in reasoning from first principles 11
3. Design-and-build projects 23
4. Using true/false quizzes to stimulate discussion 33
5. Using previously unseen calculation exercises as the basis for tutorials 43
6. Supervising experimentation in the mechanics lab 57
7. Using written assignments to promote learning 69
8. Predicting the outcome of an operation of a real mechanism 79
9. Adapting lectures to ensure that students are active, not passive 87

Conclusion 93

Appendices

1. Sixteen additional ideas for projects 97
2. Eleven additional examples of calculation exercises 115
3. Examples of questions that can be used in lab tests 123
4. The student feedback survey form developed by the author 129
5. The author's other books on basic mechanics and on teaching 135
6. Selecting the right mix of exercises to use in assessment 139
7. Entrance testing as opposed to exit testing 143

About the author 148

Introduction

The purpose of this book

For anyone in any technical capacity, the importance of having a sound knowledge of basic mechanics cannot be overstated.

Mechanics is the foundational, and probably the most important subject for an aspiring engineer to grasp. It is the science that deals with how solids, liquids and gases interact with forces, the rules that govern the equilibrium and movement of objects, and the ways that energy is transferred.

A course in basic mechanics should ideally:

- Result an a high level of motivation in students, and
- Imbue students with the kind of attitude and a way of thinking that enables them to do engineering.

These qualities arise from participation, intrigue, challenge and accomplishment. This book aims to give instructors practical advice on how to structure learning activities that optimise the chances of such outcomes.

Here will be found descriptions and examples of nine effective types of learning activity that can be used in teaching the subject of basic engineering mechanics.

Some of these types of learning activity are already widely used, although not always optimally. Some are less well-known, but deserve to be used more often. The emphasis here is on tailoring learning activities to ensure student involvement, get them thinking, and be as practical as possible.

What is meant by a 'learning activity'?

All meaningful learning, irrespective of the subject, is accomplished by *doing* something, not from just hearing about it, or seeing visuals, or attempting to arrange second-hand information, as when writing an assignment based on references.

You might 'get the picture' by having a topic explained, but until you actually do something with your mind and hands, and see the results, you won't have internalised that knowledge.

This observation is particularly relevant to fields of endeavour that require the development of a specific type of thinking approach, through experience. Like engineering.

What is the difference between a teaching activity and a learning activity?

There is no hard and fast difference, because, obviously, an instructor does something called 'teaching' that is supposed to result in 'learning'. The difference is only discernable by examining the extent to which learning actually occurs, as a result of the activity.

We tend to think of teaching activities as those during which there is a tendency for the instructor to do a lot and the students relatively little. A standard teaching lecture is one such event.

Conversely, a learning activity is one in which the students do a lot, and the instructor relatively little, as far as may be observed *during* the event. However, what happens *before* the event is critical: if students are to achieve an appropriate type and level of activity, the instructor needs to have prepared for it very well.

A 'learning activity' as described here, is one that sets a focused task for students to tackle. Performing the task ought to result in them getting to understand and be able to apply the principles that they need to learn.

In a well-designed learning activity, students will:

- Learn by doing something practical,
- Take part in discussion and cooperation, directed at solving specific problems,
- Obtain feedback from peers about the way they interpret the subject, and
- Obtain direct feedback about their grasp of the subject from the way their hands-on efforts turn out.

Most of the learning activities described in this book can also be used by instructors to assess how much the students know.

Why isn't conventional engineering teaching based primarily on practical learning activities?

In the field of academia that is devoted to the study of teaching and learning, a lot of lip-service is paid to the ideal of 'hands-on' learning, but that ideal is rarely achieved, even in subjects that are essentially practical in nature. Why?

There are several reasons:

- Some university and college departments rely too much on lecturing as a staple teaching method: they expect instructors to lecture, and they timetable for that. Essentially, if you are a lecturer, you have 250 students expecting something to happen on Monday, first period, so you had better do something with them, and the easiest option (for you) is to lecture.
- Study material is inevitably geared toward the skills valued in answering examination questions. If the assessment process in the institution calls for rote learning and over-abstracted approaches to the subject, the teaching will follow suit. Assessment processes tend to incline in this direction for the convenience of examiners, as a result of some of the other factors described below.
- Many instructors, personally, in the course of doing their jobs as academics, operate on an abstract plane where mathematics, statistics and computation make up most of their sense of reality. Some of them might be out of touch with the practical level of engagement needed in engineering projects. Some of them have *always* operated on the abstract plane, and find it difficult to imagine why not everyone thinks like they do. Some students, in fact, most, in my experience, need to have hands-on experiences that will enable them to grasp the abstractions used in their discipline.
- Many instructors in universities are obliged, of necessity, to focus most of their attention on research and publishing, so don't have time or energy to devote to developing new learning activities or exercises. They also don't relish overseeing the conduction of time-consuming activities from which their students might benefit.
- Even if they had the creativity and energy to redirect their teaching toward making use of hands-on activities, they wouldn't get recognition or advancement for it. The reason for this is simple: educational institutions cannot reach general agreement on what constitutes 'good' teaching.
- Many teaching institutions have become waylaid by the excessive use of information technology in their teaching process. Students spend huge amounts of time interacting with computers, attending to learning management programmes and to so-called 'interactive learning programmes' by means of which they experience modelling rather than realities. While institutions might claim that this trend is inevitable, due to large class sizes and events like the Covid 19 pandemic, in a large part the trend is fuelled by the lure of the latest gadgets. As in: 'Dear instructor, here's another tool that you didn't know you needed. It will change your life!' Yes, but for what gain? It seems that, in the quest to remain at the cutting edge of the discipline, educational leaders overlook the fact that engineers have been trained on the job quite adequately for thousands of years before the advent of computers.
- Classes are often too large for the logistics of practical exercises to be easily arranged. Class sizes get this way because universities have morphed into income-generating bureaucracies. Despite their grandiose mission statements, what really drives them is 'bums on seats'.

- Many engineering departments face a demand from upper management to produce high levels of 'throughput', namely to show that a high percentage of students pass. This places the focus of the 'teaching' on passing rather than on learning. In order to pass, a student has only to play the academic game well enough to succeed. As in all games, participants look for the most efficient way of achieving the desired result, which very often means doing just enough to pass. Taking that approach to one's studies does not result in meaningful learning, or significant retention. However, it is not only students who are affected by the requirement for high throughput. Instructors tend to avoid using exercises or activities that place high demands on students, for the very reason that the results of assessing student performance in such activities will negatively affect the throughput figures.

It is time to get real.

The two categories of learning activity described in this book

1. Traditional 'teaching' activities

In engineering schools, the traditional types of activity most often used in teaching basic mechanics are:

- Lectures explaining the theory
- Demonstration of and practice in tackling calculation-type questions,
- Lab experiments,
- Written assignments, and
- Tutorials.

The way these activities are used in practice is not always optimal. In this book I analyse the reasons for this, and show how to improve on the effectiveness of these traditional activities, in order to swing the emphasis from teaching to learning.

2. Activities that put emphasis on student participation

These are activities that can be used in teaching, which definitely involve students to a greater extent, and consequently result in more meaningful learning. They include:

- Reasoning from first principles,
- Carrying out design-and-build projects,
- Getting feedback from true/false tests, and
- Predicting the outcome of an operation of a real mechanism.

This book provides a brief analysis of each of the above types of activity, showing the pros and cons of using each one as a learning opportunity, and, where appropriate, as an assessment tool. With each type of activity that is described, a small number of examples and suitable exercises are provided.

All the exercises used as examples in the present book and in the author's other books on mechanics *(described in Appendix 5)* have been devised by the author. Most of them are entirely original, in terms of how questions are phrased and illustrated, and in the choice of practical applications to which the exercises relate.

Some of the exercises are inevitably variations on those that might be encountered in other tertiary courses in basic mechanics. Any similarity between some of these exercises and others that the reader may know about, is simply due to the nature of what needs to be learnt about the subject matter.

Instructors are welcome to use these exercises any way they like. It would be appreciated if acknowledgement is made to the source of the material.

The *methods* of solving these exercises are not given here, but are set out clearly in the author's series 'Basic Engineering Mechanics Explained'.

The topics in engineering mechanics that provide examples for the learning activities in this book:

The subject content covered in the exercises in this book complies with the bulk of the topics found in the first and second semester syllabi of most engineering colleges and universities. Many of these topics are also taught in high schools that prepare students for entering engineering courses. The topics are:

1. Concepts, quantities, principles and laws
2. Working with numbers in engineering
3. Forces: components, resultants and equilibrium of particles
4. Force moments, torque and equilibrium of rigid bodies
5. Centres of mass, centres of gravity and centroids
6. Forces in the members of trusses under load
7. Friction between dry sliding surfaces
8. Buoyancy
9. Linear motion with uniform acceleration
10. Motion influenced by gravity: vertical and projectile motion
11. Rotary motion
12. Work, energy and power
13. Simple lifting machines
14. Inertia in linearly accelerating systems

15. Linear momentum and impulse
16. Relative velocity
17. Centrifugal and centripetal forces
18. Rotational inertia
19. Rotational inertia combined with linear inertia in accelerating systems
20. Kinetic energy of rotation and angular momentum
21. Simple harmonic motion, and
22. Basic vehicle dynamics

Chapter 1

Running effective tutorials

Properly-run tutorials are probably the most effective form of learning activity in any subject, including engineering. Unfortunately, many instructors have never been shown the finer points of what makes a tutorial effective.

The designs of activities called 'tutorials' vary according to the traditions of the departments where they are used.

The author conducted research into the tutorials run by 33 university departments at the University of Cape Town, to see which tutorial designs worked, and why. The results of this research are described in detail in his book 'Tutorials That Work, a guide to running effective tutorials' *(currently out of print, though copies may be obtained from the author)*.

What makes a tutorial effective?

The essence of the findings of the above research can be summed up by this definition:

An effective tutorial is an event in which students, through discussion and cooperation, strive towards answering a specific question (or questions) after examining some stimulus material presented to them on **the occasion (not** *prior to* **the occasion). The stimulus material could be an object, device, diagram, photograph, video clip, calculation exercise, quiz or a piece of writing.**

Let's call this effective type of tutorial structure 'type **A**'.

Here is a description of a real 'type **A**' tutorial that was observed in the course of this research:

An archaeology tutor brought to the class a set of bottles in current use, for a variety of products. She asked the students in groups of three to examine these bottles and make notes about which bottle features were specifically suited to the respective products that each bottle contained, according to the label. When they had made their notes, the modern bottles were removed, and she produced some very old glass bottles that had been unearthed, that had long since been without labels. She asked them to try to determine what sort of product might have been contained in each respective bottle in its heyday.

This tutorial was successful in that it contained:

1. immediate and tangible stimulus material,
2. a specific task,
3. excellent opportunity for purposeful discussion by everyone present,
4. the necessity for participants to go beyond comprehension to synthesis.

Many university departments had other types of tutorial designs which were found ineffective or tedious by students. The reader may be familiar with these common examples:

1. **Type B**: Some instructors give out a question sheet in advance, which they call a 'tutorial'. That sheet contains calculation-type exercises to be done by the students in their own time. The subsequent face-to-face encounter is supposed to be an occasion for students to compare their answers and get guidance from a tutor on how to answer the questions. Despite its widespread use in the sciences and engineering, this lesson-plan is not popular with students or tutors. The actual contact period is downright boring. Basically, this tutorial design has four big shortcomings: (a) some students don't complete the work in advance, so they can't make a useful contribution to any discussion (b) too much of a student's time is wasted while the tutor is explaining something that you already know, (c) the tutor doesn't always get around to addressing all the questions you may have, and (d) there is very little opportunity to discuss anything with other students, because there is no discussion task set, and no student has any idea of how much the others know or might be interested in helping him or her. When I was an engineering student, I often reeled away after being present at such 'tutorials', completely dissatisfied with the way my time had been spent.
2. **Type C**: Some lecturers issue reading homework, and expect students to come together to 'discuss' the material under the supervision of a tutor, with some prompting from the tutor. This arrangement doesn't work either, because many students don't do the required reading. A lot of students bank on using the tutorial to deduce what they were supposed to have learnt by picking the brains of those who *have* done the reading. In the 'discussions' that take place, the centre of attention is held by one or two students who are either very argumentative or inclined to show off, and there is no real interaction with those on the periphery. There is also no specific task set, so the 'discussion' runs out when no-one has anything further to contribute.

The reason why the lesson-plan of a type **A** tutorial is effective, is that you can only take an active part in 'discussing' something that you have engaged with, *while the engagement is fresh in your mind*. Also, you only have an incentive for active discussion if there is a problem-solving purpose to the discussion and everyone starts on the same page.

In my research on that campus, in all instances in which type **A** tutorials were observed (by trained research assistants), students reported enjoying the event, and looking forward to the next tutorial. Students attending tutorials with the structure of types **B** and **C**, did neither.

In what way does a tutorial create a learning opportunity?

You learn by developing your own way of understanding the subject in hand. To make this possible requires that students articulate aloud how they personally interpret the stimulus material, so that they can sense from their peers' responses to what they have said, where their own thinking can be substantiated, and, conversely, where it might need modifying.

However, a tutorial has to *set up* an opportunity for meaningful discussion, or else the advantages of putting people together in groups will not be realised.

The ideal group size for tutorials

In order to have a chance to express one's thinking aloud, the discussion group has to be small enough for everyone to have a say. Having three members is ideal, four is acceptable, but with five or more there will always be some people who don't get a word in edgewise. So, if you have a larger group, it is sensible to divide it into sub-groups.

If needed, later in the proceedings each sub-group can contribute its solution to the tutorial task in a plenary involving all the sub-groups.

What if some students don't 'click' with others?

At the beginning of a term, it is useful to allocate students to completely different sets of peers for three or four group activities, so that they get to meet others on the course, and experience trying to work with a range of different people.

In the course of a tutorial sub-group discussion, each participant will sense the way the others think. There will be a natural process of gravitating towards cooperating with those peers that you feel take a similar approach to your own, or whose minds work in harmony with the way that yours does.

After these first few events, there is no point in re-allocating students to yet more 'new' groups. By this time, students will have formed alliances that they would prefer to keep. It is absolutely natural, in all aspects of one's life, to find it easier to collaborate with some people rather than others.

There is a good reason not to break up these alliances, namely: if you find people with whom you can work, you can get some work done.

What is the role of peer responses to what a student says?

Peer responses constitute feedback. For feedback to be of any use, it needs to be constructive, not destructive. Think about the respective psychological effects on your own further participation in a discussion when you are on the receiving end of statements like the following:

a. Now, that's interesting. Why do you say that?
b. You have a point there, but how would you take X into account?.
c. You say the friction is caused by X. Are you sure? What about Y?
d. No, it doesn't work like that!
e. That's a stupid idea!
f. Don't be a *(descriptive word)* idiot!

Every participant in a tutorial discussion should be made aware of the simple rules for giving constructive feedback. The essence of good feedback is to show interest, not approval or disapproval. Students and instructors would get more out of tutorials if they understood how to give constructive feedback. My book 'Motivating People to Learn… and Teachers to Teach' *(see Appendix 5)* devotes an entire chapter to this topic.

Can one expect all students to respond equally productively to peer feedback?

No. Everyone is different. Quite aside from differences in intellectual abilities, and differing styles of thinking, there is a variety of psychological factors that affect an individual student's approach to processing information during a discussion. Namely that:

- Some people are more open to changing their mindsets in response to new information, others less so.
- Some are more argumentative, others more able to reason without high passion.
- Some people give up easily when attempting to learn something that they find

difficult, while others are more patient and persistent.
- The same student can show varying amounts of enthusiasm for different topics, depending on how relevant the topic appears to them and the way they are expected to engage with it.
- Some people doggedly don't care what others think, imagining that the only valid viewpoint is their own. Anyone who disagrees with them is viewed with disdain.
- Some people are hugely sensitive to what they perceive as a slight to their character or abilities, while others are able to brush aside any such inferences and concentrate on the issue to be solved.

In this respect, the discussion that takes place in a tutorial is a microcosm of social life in general. So, no: not all participants in a tutorial will take in the feedback they receive to an optimal extent. However, none of the students would benefit from feedback, if there was no feedback to be had.

Can tutorials be used as assessment events?

A tutorial's most significant contribution to the assessment process is to assist students to assess their own knowledge, by comparing their thinking with that of their peers.

If you want to use a tutorial as an assessment event, you have to assess what students have learnt from the experience, not how well they prepared for it.

It is not advisable to ask for work done before the tutorial to be handed in during or after the tutorial itself: there are too many ways in which that work might not be an accurate indicator of the effort put in by students.

Reliable assessment of what they have learnt is only possible if you issue a separate test held under test conditions, subsequent to the discussion.

If you make a regular practice of issuing a short test following a tutorial (marked and contributing to a year-mark) this could be an additional incentive for students to attend tutorials.

Can one assess participation levels in tutorials?

No. Some people are adept at giving the impression that they know what's going on, when they don't, actually. They could just be windbags, extrovert performers, or trying to impress the tutor.

In a group 'discussion' attended by twenty students seated in a large circle, some people will be disinclined to participate. Those who say very little might be any of the following:

- Introverts who look upon the proceedings from the sidelines, quietly gleaning whatever they need to know.
- Uninformed and therefore incapable of participating
- Not confident at speaking to large groups
- Good listeners

In a large group discussion, there is no relation between the amount of talking you do and the amount of learning you do.

If the sub-groups are small enough, everybody *will* participate.

What kinds of exercise can be used as the basis of a successful tutorial?

Almost any exercise that can be devised around the subject of mechanics (and any subject, actually) can be turned into a type **A** tutorial, as long as there is newly presented stimulus material and a specified discussion task around it.

As stimulus material, one could use, for example:

- One or more calculation questions, provided they are previously unseen,
- An existing piece of machinery that can be handled and reverse engineered,
- A true/false quiz to stimulate discussion,
- A brief for a design project, an approach to which needs to be brainstormed, or
- A sample lab report to be marked by the students, then analysed and improved upon.

Examples of tutorial 'lesson plans'

Only two examples are presented here, since instructors can apply the structure of a 'type-**A** tutorial' to whatever material they are teaching. The reader will find many possibilities for tutorial material in several of the subsequent chapters.

Also, many more ideas for tutorial questions can be found in the author's series of three volumes: 'Basic Engineering Mechanics Explained' *(see Appendix 5)*.

Note: It is not always possible to provide 'correct' answers for every tutorial, because for some types of task, the discussion can go in a multitude of directions, depending on who is participating and what they know. Tutorial 1 (described below) is one such.

Some tutorials can be designed to specify particular goals, which can be arrived at by calculations performed by the sub-group participants. Tutorial 2 below is an example. For this one, numerical answers are provided.

Tutorial example 1: Analysing a Student Lab Report

The accompanying report (*which students would be shown, but which is not given here*) is one student's description of an experiment carried out by his group, in which the group:

a. Determined the radius of gyration of a flat plate about a given axis by calculation, based on its mass distribution in relation to that axis,
b. Timed the period of small oscillations of the plate when allowed to swing as a compound pendulum, hinging about that axis,
c. Used this value for the period in the compound pendulum equation to determine the radius of gyration of the plate, and
d. Compared the results of part 'a' with those of part 'c'.

Students should read the report and answer the following questions:

1. Was the aim of the experiment described with clarity? Could you improve on the wording of the statement of the aim?
2. Does the report give sufficient detail for one to repeat the experiment with confidence that it could be done exactly as reported?
3. How reliable do you find the description of the way in which the measurements were taken?
4. Could you detect errors in the experimental method that do not seem to have been accounted for in the report?
5. The radius of gyration of the plate about the given axis was determined by two methods. Which of these methods produces a more reliable value? Why?
6. Explain how you would determine the percentage difference between the values produced in step 'a' and step 'c'.
7. What percentage difference between the results of the two processes would you consider acceptable to conclude that these two processes do in fact, produce the same result?
8. How many repeats of the experiment would you consider necessary in order to claim that both methods produce an equivalent result?
9. For more confidence in the wider applicability of the results of this experiment, would this experiment need to be carried out using differently shaped plates? If so, which shapes of plate would be suitable, and which not?

Tutorial example 2: The Grindstone

The grinding wheel of an old-fashioned hand-cranked (geared) grindstone consists of a stone of density 2700 kg/m³, diameter 788 mm and thickness 104 mm. The stone is brought to a speed of 60 r/min and then allowed to freewheel until it comes to rest, which takes 1 minute and 46 seconds.

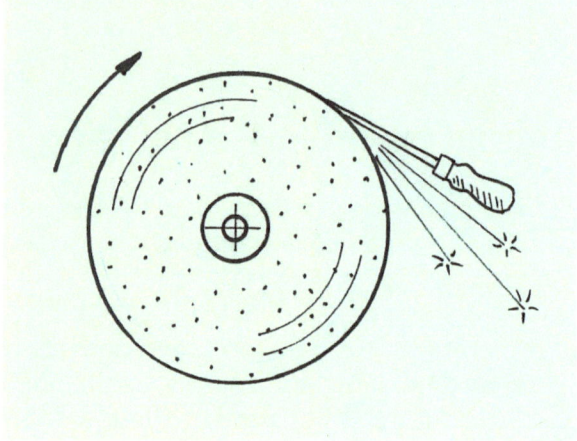

a. Determine the mass moment of inertia of the grindstone wheel. (Ignore the central hole and the axle through this hole: just consider it as a solid disc.)	
b. Determine the frictional torque in the bearings.	

Now, the stone is again accelerated, and is used to sharpen a tool, at a steady speed of 160 r/min. The tool is pressed against the surface of the stone with a radial force of 9 N. The coefficient of friction between tool and stone is 0.4.

c. If the person powering the grindstone stops turning the crank, but the tool is kept pressed against the stone with the same constant force, how long will the stone take to come to rest?	
d. What is the angular momentum of this stone at 160 r/min?	

e. What value of braking torque would be required to bring the stone to rest from 160 r/min, in 6 seconds? (Without the tool against the surface.)	
f. When turning at 160 r/min, what is the relative velocity of the stone's grinding surface, as seen from the tool tip?	
g. What steady power input is required from the person turning the crank while this tool is being sharpened?	

For those who wish to issue or tackle this tutorial, the following answers are provided:

10.63 kg.m² 0.6300 Nm 86.95 sec 178.1 kg.m²/s 29.05 Nm 6.601 m/s 34.32 W

Chapter 2

Exercises in reasoning from first principles

This kind of exercise requires students to assess a given practical mechanical situation without all the necessary data for a 'solution' being provided.

They have to decide for themselves which principles apply, in order to answer the questions. The reasoning they need to use does not necessarily have to make use of calculations, but has to be logical and practical.

Example:

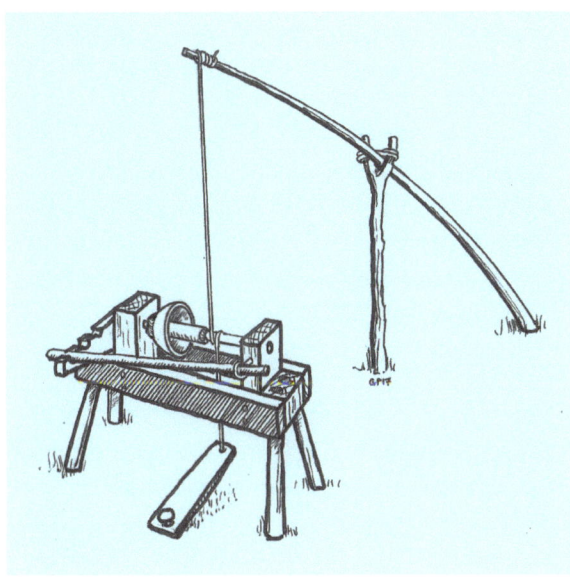

What is this device? What is it being used for in this particular illustration? How does it work? Can you identify the function of each component? Which principles of mechanics affect the operation of this device? In normal operation, is there any likelihood of the 'table' toppling over? Why, or why not? Which properties are important for the materials of the various components to have?

Although no mathematical relationships are being invoked, it is clear that such a device is mechanical, and must therefore operate according to the principles of mechanics. Can the student speculate firstly on how the mechanism works, and secondly, discern which principles of mechanics apply?

Why make use of such exercises?

To appreciate why one is learning principles, one needs to learn to see those principles applied in a real context. Too many standard questions used in teaching engineering mechanics presume that students would be aware of the relation between the theory they have absorbed and the physical context, when, in fact, they are not. Inexperienced students can look at a real mechanism, even understand how it works, but not see any connection between it and the equations they are expected to memorise.

Is it necessary for instructors to teach everything about mechanics in a mathematical way?

This author has encountered the (usually unspoken) assumption by many instructors that engineering consists of applying mathematics to every conceivable situation.

Such an assumption is problematic, for several reasons:

- In practical day-to-day engineering, precise mathematical reasoning is far less necessary than it is in engineering classrooms. While you are a student, you may easily get the impression that mathematics is indispensable for solving every problem that an engineer encounters. However, once you start your career you realise that decisions are more frequently made on the basis of pragmatic considerations than on the basis of advanced calculations. You begin to find that experience is more useful to you than mathematics. It is not that your mathematical and scientific training was a waste of time: rather, it provides you with a backdrop of informed (but by now partially intuitive) appreciation for quantifying the variables you encounter. Detailed mathematical analyses are sometimes necessary (more so in some industries than in others), but they are always only *part* of the solution to any given problem. The biggest requirement of a 'solution' to any engineering problem is that it should be practical.

- With some work experience, as a young engineer you also begin to realise that real life engineering problems are never neatly and precisely defined for you, the way they were in class. You will need to navigate your way among a confusing array of constraints that complicate any approach to defining a problem, before you can make an attempt to solve it. These constraints include considerations of cost, availability of resources, awareness of the environmental consequences and the political and human needs of both the clients and the population at large. In other words, you will have a 'big picture' to look at, and against its backdrop, you have to define the problem for yourself, because there will not be an instructor at hand to define it for you.

It follows that it is completely justified and highly meaningful for an instructor to include opportunities for students to show their mettle in appraising and solving everyday problems by reasoning from first principles.

To be able to reason like this calls on 'common sense', which is becoming increasingly less common in the population. To exhibit common sense in the area of mechanics requires a certain amount of experience in one's childhood of having played with real materials, like sticks, stones, ropes and wheels. It also requires sufficient interest and focus to pay close attention to what one is observing.

One factor that has contributed hugely to the diminishing of common sense in the population is technology itself. By having one's life centred around computer-driven possibilities, young people develop what I call a 'menu-mind'. Their life choices are increasingly confined to choosing options from a menu that someone else has designed. When you take them away from their electronic devices, and face them with a completely open physical situation to put their minds to, many of them don't cope so well. This is the paradox of clever engineering: the smarter our technology gets, the dumber our descendants become. They may be smart in the sense of learning to operate the technology, but are they smart enough to appraise complex situations, and think for themselves?

Another vital reason for setting up opportunities for students to engage with a real physical problem and come to their own conclusions about it, is that doing this is actually enjoyable for the student, and therefore motivational. It is far more motivating than sitting listening to dry-as-dust lectures. From a learning point of view, it is more useful even than listening to well-delivered, interesting lectures.

The point of studying engineering is to develop the capacity to solve problems. That is more important than trying to absorb the formalised knowledge-structures of one's mentors. In the long run, every individual is going to develop his or her own knowledge-structure, anyway. That process is powerfully influenced by what you find interesting and what you find boring.

The contact activities that students enjoy most of all are those that give them opportunities to use their own judgement, no matter how limited their experience is.

There are various ways in which an exercise might require reasoning from first principles. These include:

- **Analysing a device by examining it or a drawing or a photo of it, and**
- **Assignments that present descriptions of hypothetical practical situations in which there is a problem that requires estimation, reasoning and creativity.**

In both of these types of activity, the opportunity for cooperation and meaningful discussion among participants is huge. So, students can learn from bouncing ideas off one another, which is the natural way of building a knowledge structure. Due to this feature, many such assignments are suitable to be used as the basis for tutorials.

Analysing a real device or a design of a device
(also known as reverse engineering)

Reverse engineering is the process of figuring out the thinking behind the creation of a design, by examining and discussing the outcome of that design.

Doing reverse engineering gives students experience in the design process. They benefit from the opportunity to identify shortcomings in someone else's design, so that they can avoid perpetuating similar mistakes. They also discover which design features are admirable, and thus worth remembering.

Basically, this is the process by means of which humans have *always* developed anything physical. You can't re-invent the wheel, but you *can* improve on the wheels you have tried to use. Doing reverse engineering gets you thinking about which principles are involved in the particular design you are examining, which helps you to become conversant with those principles.

Some might feel that reverse engineering assignments are more relevant to a design course than to a mechanics course. However, mechanical devices only work because they have successfully made use of the principles of mechanics. It is highly relevant to a course in mechanics, for students to detect which principles apply to a given mechanism, and to appreciate how to apply those principles optimally.

For an exercise in reverse engineering, students should be provided with something tangible to examine, and prompted to answer questions about it. Basically, such exercises are ideal type-A tutorials.

The item that the students have to engage with could be a real device , or it could be a drawing or photograph. The subject matter could come from any one of the following general categories:

- a purely mechanical machine designed and built by other students, such as a small trebuchet,
- a mechanism from an industrial or farming museum, (such as a book press or seed-planter),
- an antique kitchen utensil (such as an apple peeler),
- a part of an unfamiliar motor, engine or pump,
- a maritime pulley block or capstan,

- a utensil not found in our culture, such as a Polynesian weapon,
- a print depicting a centuries-old mechanism such as a medieval water-pump, or
- a diagram of a mechanism recorded by illustrators from ages past (such as the drawings of Leonardo da Vinci).

Typical questions to ask

- What is this object?
- What is it made of?
- What does it do?
- How does it work?
- How do you think it was made?
- How old do you think it is?
- Do you find anything admirable about the design?
- Is there any way you could improve on it, given the intended function?
- Which principles of mechanics have been used in this design?

When the students have finished examining and discussing the item, in order to assess what they have learnt from the exercise, each student could be asked to do one of the following:

a. Individually write down their answers to the given questions on a piece of paper, with sketches if appropriate, and possibly submit that before the end of the session, or
b. Use the information thus gained to come up with an improved design, or
c. Build their own similar device, the performance of which is going to be tested, or
d. Take a subsequent written test that presents them with the same or similar questions about this design.

An example of a reverse engineering exercise used successfully by this author

In my first ever teaching trimester, I had been disappointed by the level of understanding shown by my students about the principles of operation of simple lifting machines. I thought that I had explained the basics quite well in my lectures, but it turned out that the students were very hazy about these principles when asked about them in a subsequent test.

So, with the next intake of students, before telling them anything about simple lifting machines, I decided to let them get some experience so *they* could explain those principles to *me*.

I arranged for all the types of simple lifting machines we possessed in the lab to be fixed to the walls or an overhead beam. Students were allocated to a machine in groups of four, and instructed to play with the machine for 15 minutes.

Then, each group had a turn to tell the rest of the class how their machine worked and what they could deduce about *why* it worked like that. The lab was a real buzz that day. Students got busy discussing and trying out the equipment and having a lot of fun. Best of all, it was clear that every group had easily managed to figure out the way their particular machine worked. The following day, when I summarised the principles that applied to the topic, due to their experience they were able to ask me sensible questions and understood my answers. It was a complete success. Besides, they nailed the subsequent test, because context and experience had preceded theorising.

Some typical exercises for analysing a device

The following exercises are not all of the same level of complexity, and they don't follow any particular sequence. Some of them can be used at the early stage of a course in mechanics, others after more exposure. They are merely offered as an illustration of the range of exercises that are possible.

Three examples are provided here. More are provided in Appendix 1.

Exercise 2.1 Analysing a machine built by other students.

Examine this photograph of a student-built machine. Discuss in your group:

a. What would be the purpose of this machine?
b. How is it supposed to work?
c. What do you think were the instructor's constraints on the materials allowed to be used in this project?
d. In what form is the energy being stored?
e. Why is his foot on the base of the machine? Which principle of mechanics is making this necessary?
f. What is the purpose of the cord that is lying loosely over his shoe?
g. Which features of his design do you think look like good ideas?
h. What improvements would you make if you were to use this design concept?
i. How does this machine differ from the one seen in the background?
j. Which of the two designs would you prefer, and why?

Exercise 2.2 Ritchel's Dirigible

In 1878, Prof Charles F. Ritchel designed and built a dirigible which flew at an exhibition in Philadelphia, piloted at first for an indoor flight by a young woman, Mabel Harrington, and subsequently outdoors by a teenage boy, Mark Quinlan. Ritchel couldn't operate it himself, as a grown man's weight would have been too much for the amount of buoyancy provided by the hydrogen-filled blimp.

The illustration here (from the cover of Harper's Weekly of July 13, 1878) shows his 'flying bicycle' in flight. The artist fancifully put Ritchel at the controls, typical of media distortion.

Some details: the blimp was a cylinder of fine linen with a rubber coating, containing hydrogen. It was 25 ft long and 13 ft in diameter, and weighed 66 pounds.
The frame was made of hollow brass tubing. The total weight of the

machine was 112 lb. The boy pilot weighed 96 lb.

Power for the propeller was provided by the operator turning a hand crank, similar to a bicycle pedal crank, but with a larger sprocket. In order to provide directional control, the propeller axis could be turned by a foot-operated control mechanism. The first outdoor flight in Hartford, Connecticut, reportedly went out over the Connecticut River and back, landing at the starting point.

Some questions:

- Check on the buoyancy to be expected from this blimp. Would it have been sufficient to keep the machine and pilot airborne?
- Do you think the members connecting the blimp to the lower frame were solid or flexible? Why? Which option would have been preferable?
- Nowadays, what material could we use in place of the rubberised linen? Substantiate your suggestion.
- Would it be advantageous to replace the brass tubing with another material? Which? And why?
- How much of a weight saving could be achieved with the materials you suggest?
- How do you think the mechanism that steered the propeller worked?
- How much power could a teenager exert in this fashion?
- What sort of speed do you think could be achieved using this means of propulsion? Can this speed be estimated by a plausible calculation?
- How would you control the altitude of such a craft?
- Would a different shape of blimp incur less air resistance? What shape would you suggest?
- There are no safety features evident in this drawing. If you were to build one today, what safety features would you incorporate? *Apparently on a subsequent flight, at an altitude of about 200 ft, the propeller gearing jammed, so the craft could not be steered. The boy pilot inched forward along the gantry, tied one ankle and his left wrist to the frame, then, hanging from the frame, he unjammed the gearing with his pocket knife, before working his way back to the seat and continuing.*

Exercise 2.3 Medieval Crane

This illustration is a representation of a medieval crane used to lift and place building materials. One, or sometimes two men were stationed inside the treadmill **C** which they walked inside to provide power to wind the rope onto or off the drum. On larger cranes, more than two men would be needed to work the treadmill.

Answer these questions:

a. If a man on the treadmill needed to avoid his head hitting the drum axle, estimate the diameter of the treadmill.
b. Using your estimate of the treadmill diameter, estimate to scale the dimension **AB** of this crane boom, and all other significant dimensions.
c. What looks to be the maximum height to which a load could be raised by this crane as it appears here?
d. Estimate the combined weight of the components of the boom structure suspended to the left of the point where the boom is attached to mast **D**. Include that part of the boom to the left of the mast.

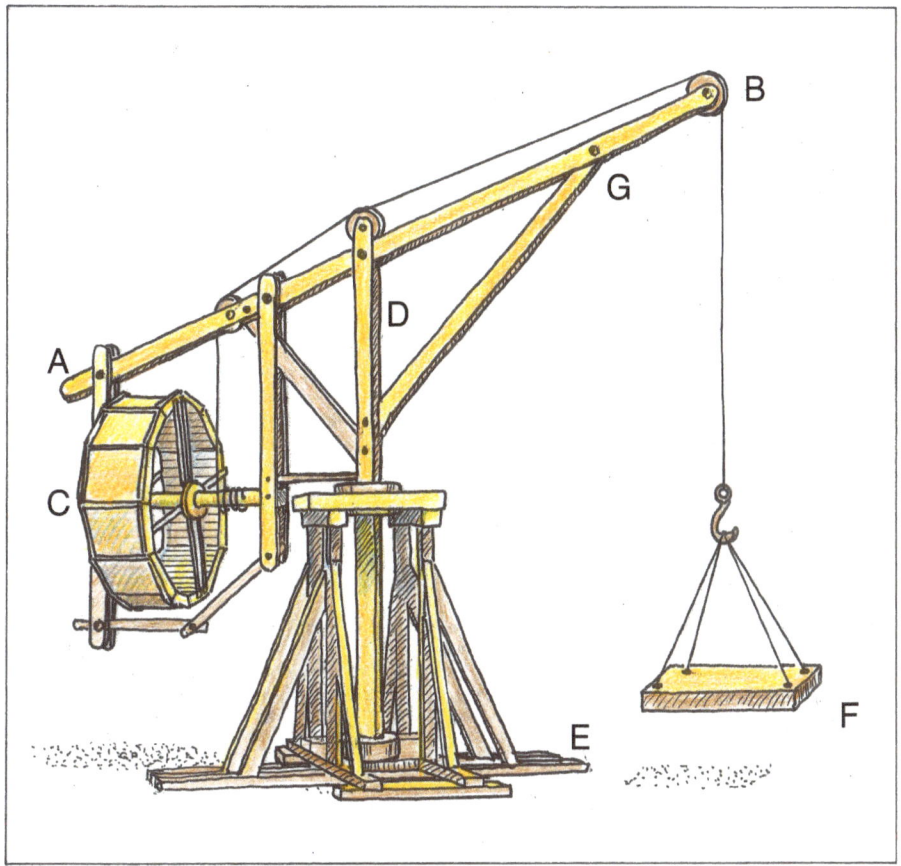

e. In order for there to be no bending moment on the mast at the point where it emerges from the base, what needs to be the weight of the load platform **F** with its load?
f. What would be an appropriate maximum load rating for the rope?
g. Do the diameters of the three sheaves affect the velocity ratio of this crane? If so, how?

h. Name all the factors that negatively affect the efficiency of this crane.
i. If the rope should break in use, what will happen to the treadmill?
j. Presuming that nothing breaks, what would be the approximate value of the load that would cause the entire crane to tip forward about point E?
k. At which points on the crane would you expect the bending stress to be high? Why? What would you do to the design to minimise these points of stress concentration?
l. Estimate the turning moment that can be supplied by two men exerting their full weight 300 mm forward of the lowest point of the treadmill.
m. If this value of turning moment is exerted, and the winding drum has diameter 200 mm, what would be the maximum load that could be raised? For this calculation, ignore all sources of friction.

Assignments that present descriptions of hypothetical practical situations in which there is a problem that requires estimation, reasoning and creativity

This kind of assignment presents a hypothetical situation that might require building a device in order to solve a problem. Students won't *necessarily* have to tackle the proposed construction tasks physically, but will need to think about them *as if* the assignments were going to require physical engagement.

If desired, some of these assignments could also be escalated to becoming design-and-build projects, provided the logistics can be taken care of.

The project is assigned in a brief, which includes details of *some* of the materials that would be issued for this project, and asks questions about what else might be needed to tackle such a project in reality, and how students propose to go about tackling such a project.

A paragraph something like the following could precede the project brief:

Consider the project described in the following brief (in bold print below). You will not actually have to carry out the brief, but you should answer the questions that follow it, as if you were going to carry it out. In order to answer the questions, you need to spend some time alone this week, thinking about the project and making some notes and sketches, and then share your ideas at the subsequent small-group discussion that is scheduled for next week. Subsequent to that discussion, there will be a test on this assignment.

One such project brief is given below. Three others are presented in Appendix 1.

Exercise 2.4   The Bridge

A team of 6 people is required to build a bridge across a strongly-flowing stream that is 5 m wide, 3 m deep and icy cold. The water level is only 0.5 m below the banks, which are near vertical, and consist of soil, held together by grass. This bridge has to allow 20 adults with backpacks, each of total mass 100 kg, to cross it, with two of them being on the bridge at any time during the crossing, one behind the other. Available material: 12 poles of green wood with the bark still on it, diameter between 50 and 60 mm, each 3.6 m long, 20 poles which are of 2 m length, of similar fresh-cut wood, diameter 35 to 45 mm. You would also have 50 m of 12 mm diameter nylon rope and 200 m of 8 mm diameter hemp cord. Tools: one hand-axe, one bow-type pruning saw, one spade, 2 clasp knives, one broad flat chisel and one 2 kg sledgehammer. The bridge has to remain in place and function, without maintenance, for at least two weeks. You have one week's notice to prepare yourself for the project. Materials will only be issued on site, on the day of testing. On that day, you will not be permitted to bring to the site any paper or electronic devices, or any tools besides those issued to you. You may only work from the near bank of the stream, until your bridge is sufficiently advanced to convey team members to the opposite bank, without them going through the water.

Now, answer the following questions:

1. Where would you look for design ideas?
2. Would you consider only one design for this bridge?
3. Would you build a small-scale model first? Why, or why not?
4. What do you think is best: to build the bridge first and try to manipulate it across the stream, or to build it progressively from a securely anchored point on the near bank?
5. How will you cooperate on the day to accomplish your build without access to any diagrams or references?
6. Do you think you need a team leader, or could you do this democratically?
7. Which skills, if any, would you need to learn before tackling this project?
8. What kind of protective clothing might you need to wear during the build?
9. If you were allowed to ask for one more item of material (not a tool), costing no more than £10 (say $15), to help you complete this project, what would that item be?
10. What are the nuisance factors you are likely to encounter?
11. How would you deal with each of these nuisance factors?
12. If you can think of other questions to ask that would affect your approach to the work, write these questions down.
13. Which mechanical principles, laws and concepts will you apply in your approach?

Basing an assessment on this category of exercise

To use one of these exercises for assessment would require there to be a written test held subsequent to the completion of the activity. This test would probe for what each individual student has learnt from participating in the discussion activity.

Clearly, if an exercise of this type has been taken a step towards being more practical, by being converted into a design-and-build exercise, other assessment criteria would apply. Design-and-build exercises are described in Chapter 3, which follows.

Chapter 3

Design-and-build projects

Why issue design-and-build projects?

The ultimate purpose of engineering is to build things that work. Whether these things are machines, bridges, buildings, chemical processes, power generation systems, robots or consumer products, they have to *function*.

So, the best measure of achievement by engineering students is not to demonstrate that they are conversant with any theory, but to have something that they designed and built, based on the principles of the subject they are studying, that actually works.

If you suspect that projects like these are more suited to a course in engineering design than they are to a course in mechanics, consider that there is no hard and fast line that separates the subject of mechanics from the subject of design. This author has found that design-and-build projects are very effective for getting students to understand the principles of mechanics.

Design-and-build projects demand maximum involvement from students. Almost all students derive satisfaction from doing them. In participating, there is always the opportunity to be creative, and to get impartial feedback about your thinking from the way your design performs.

If you have to build something that is intended to work, you need to understand what would *make* it work. This requirement raises the game of all students of mechanics, including those who are not usually that adept at theorising.

What's involved in issuing a design-and-build project?

All design-and-build projects have to be carefully selected to be relatively safe, make use of easily-obtained materials, not require sophisticated tooling, and have a single clear objective.

It is no small matter to manage the logistics of a large group of students, each building an individual practical item, and then to assess the performance of the devices they have built. However, it can be done, and it has been done, frequently, by this author and others.

The most extreme case of difficult logistics that I experienced for a project like this occurred when, assisted by two colleagues and five senior students, we took seven hours without a break to adjudicate 394 devices, each built by an individual second-semester student. It was exhausting, but valuable from the point of view of providing an event that resulted in maximum involvement. The event was also a reliable indicator of who had applied themselves to putting the basic principles of mechanics to practical use.

The four essential requirements

This author's design-and-build projects in the subject of basic mechanics required students to conform to the following four requirements, which are described in more detail below the list:

a. The device they build must be based *exclusively* on mechanical principles.
b. Only specified inexpensive materials may be used.
c. The device must be kept within dimensional specifications.
d. The design must be intended to *maximise* some single criterion.

Expanding on these four points:

a. The projects in mechanics that I issued were deliberately designed so that students needed to make use of mechanical principles exclusively. The rules did *not* allow the use of electrical, electronic or chemical processes, or heat. The purpose of this constraint is to compel the student to look after the mechanical side of designing, and not to be distracted by other, possibly exciting technologies, which could overshadow the learning about mechanics that is supposed to be to taking place. Naturally, in their training subsequent to their first course in mechanics, combining technologies from other fields of engineering is to be encouraged.

b. Nearly always, my practical projects confined the use of materials to those more accessible to students, such as wood, paper and string. There are four reasons for this: Firstly, most students cannot afford expensive materials. Secondly, one doesn't need access to high tech tooling to work with such materials. Thirdly, having to make something work when it is constructed of inferior materials exaggerates the likelihood of breakdown, and thus alerts students to the problems they will face when working with conventional materials. For example, if you have to make a gearwheel out of wood, and see where and how it fails in use, you will appreciate all the more those design constraints that apply to metal gearwheels. Fourthly, an important consideration is the limited workshop space in a teaching institution. It is more efficient for the college if students can build their projects in their own space, with their own tools.

c. It is important to keep the device within specified dimensions. For a start, it adds a design constraint of a type that is almost always found in practical engineering. Secondly, it allows the instructor to plan the physical aspects of the testing space: to decide which venues or outdoor spaces to assign, and what kind of testing equipment will be required.

d. It is essential that the performance of a design-and-build project should be measured by a *single* criterion, so that performances can be compared between contestants. This single criterion should be the sole means of allocating a mark for the success of the design. For example, if the requirement for a student-built machine is 'to hurl a tennis ball as far as possible, given a specified energy input, then the distance hurled can be measured and constitutes the *only* indicator of success. If the requirement for a model bridge is 'to withstand the greatest load before failing', that load can be measured and the mark awarded can be made proportional to the load achieved.

Why specify only a single criterion for the success of a design-and-build project?

If a single criterion is specified, the assessor has no need to mark drawings, reports, aesthetics and the like. Some people will be alarmed at the exclusion of such criteria, because they are accustomed to applying them in the belief that students need to be competent in all of these activities. I agree, students should be capable of doing competent drawings, writing credible reports, and building something with aesthetic appeal. However, trying to include a measure of all of these in a single mark for a project just leads to problems.

My colleagues and I tried many times to spread the marks allocated to a design-and-build project over various forms of communication, in addition to the performance. We tried assessing aesthetics, robustness, economy, drawings and reports. However, after many such projects, we eventually scrapped such complexities of assessment, because:

1. The logistics of all this marking are extremely challenging for instructors, particularly with large class sizes and no paid assistance. There is a time and a place for everything, and if you need to assess 300 machines on the basis of five different criteria your workload becomes subject to the law of diminishing returns.

2. When we tried allocating marks to different criteria, it was possible for students to pass the project on the basis of marks achieved in the areas of presentation, even if their creation didn't work. That was clearly not fair. This author is all for making good presentations, because the ability to do so is a vary valuable communication tool and can lead to advancement for any employee. However, what is the point of assessing a good drawing of a useless device?
3. Similarly, why should a student write a standardised report on the project he or she did, if the time they spent on their report-writing could have been better spent on refining their design to result in a viable outcome? It would be quite a different matter, and definitely useful, to require students to write a report on the project they built, *after* it has been tested and assessed. The report then would be assessed seperately on its merits as a report, instead of providing a contributory mark to the design-and-build project.

In this author's opinion, the central purpose of a design-and-build project is to make a machine that actually does what it is supposed to do. Therefore, it should be marked with only that criterion in mind.

Testing day

The built design is tested in public on testing day. These events elicit great anticipation and excitement among students who are keen to show (or sometimes discover!) what their machine can do. Testing days are usually the only occasions on which students' work is on public display within the department.

Such events are customarily attended by several instructors, technicians, senior students and friends of the competitors. Unlike what happens during any other assessment event, on testing days students experience a sense of social identification with the department. This in itself is a bonus for all involved.

How marks are allocated

The performance of each submitted device is measured, and compared with the best-performing device from any student (or team) in the cohort. The mark for each device is awarded as in this example: In one of my design-and-build projects the brief was to hurl a golf ball as far as possible, using a specified energy input. On this occasion the best distance achieved by any group was 9.6 m. So, that group got 100% for the project. If another machine achieved a distance of only 2.4 metres, then, no matter how nice it looked, it was allocated only 2.4/9.6 of the best mark, namely 25% for the project.

The competitive nature of such projects ensures that students don't just slap together some materials like they would have done in primary school, in order to get a 'soft-heart' mark from a teacher. Since you know in advance that your mark is only as good as your machine's performance relative to that of other students' machines, you will take the trouble to do your homework on it, and make sure it is optimised before the testing day.

How students respond to these projects

This author has 14 years' worth of evidence that such projects promote many desired qualities in students. They also reveal some undesirable motives in a small minority of participants.

Students differ markedly in their approach to tackling something practical. There will always be some students who participate only grudgingly, because their entire attitude to their studies is based on putting in minimum effort wherever possible.

I always told my students: 'If you bring me your sketches or your working model before testing day, I will help you to improve on your machine by asking you some questions about your design, that could lead to you thinking about ways of modifying it.' Those who made use of this offer always benefited. However, many students didn't bother, and left their efforts to the last minute.

One one occasion, the very day after I had issued the brief for a project due to be tested in five weeks' time, a student brought a working model to show me. It was pretty good already, and needed only a small amount of refining. This young lady was on the ball, and showed all of the qualities I wanted to see in a student.

In stark contrast to her exemplary work ethic, on another occasion, a different student not only didn't even bother to make a device, but on testing day, she tried to get away with using a machine made by another student, who had put it in the trash after it had been assessed. She was not even aware that my technician had photographed every project with its maker, so we knew at once that it was 'borrowed', and therefore merited disqualification. On top of that, she couldn't even operate the device, as she had never handled it and didn't know how it worked. It was a tennis ball-throwing machine. When she was instructed to operate it, she had it back-to-front on the starting line, so the ball came out backwards. She would have been a real disgrace to the profession. She got zero for the project, and a visit to the Head of Department.

In the engineering profession, there is no place for feeling sorry for incompetence. If you come up with a design that doesn't work, at best, you will simply not be asked to tender again. Worse, you could find yourself unemployed and sued for negligence. In the extreme, your incompetence could lead to loss of life and the ruining of your career.

Some points that students need to consider when tackling a design-and-build project :

Has anyone solved this type of problem before?
What can I learn from the way others have approached this design?
Are there any mechanisms, machine components, or ideas that I can borrow from completely different contexts?
Which of the principles and laws of mechanics need to be applied in this situation?
To what extent do I need (a) any calculations and (b) accurate calculations, and when should I be content with approximations?
What are the nuisance factors that might affect the efficiency of this design?
How to make the best use of available materials, tools and processes?
What are the cost considerations?

Examples of design-and-build projects assigned by this author

Instructions for three different projects are described below. Nine additional projects of this type are described in Appendix 1.

Exercise 3.1 Tennis-ball launcher for maximum distance

Aim: design and build a machine to shoot a tennis-ball as far as possible, over level ground, on a sports field, in the open air.

The energy input to the machine must come from the muscular effort of the person who presents the machine. The energy for the throw may be stored in the flexing of wooden parts, the raising of a weight, the torsion provided by a twisted rope, or a combination of these methods.

Consult the internet for designs of historical throwing machines such as the trebuchet, the onager, and the ballista. The dimensions of your machine must be within a cube 1 m x 1 m x 1 m when ready to fire. Any machine exceeding these dimensions will be disqualified.

Materials are limited to wood, glue, string, rope, leather and nails. Hand-made metal parts may be used for the trigger mechanism, but nowhere else. Provide your own tennis-ball for practice. On testing day you will be provided with a standard tennis ball, which must be used.

On testing day, each machine will be allowed three shots. The best of these will count. If any part of your machine fails on an attempted shot, you lose that shot. The performance index = (distance achieved ÷ mass of machine). The machine with the best performance index will be allocated 100% for the project. All other machines will be allocated a mark in proportion to this. For example, if the best machine threw 48 m and weighed 4 kg, its perfomance index would be 12. If your machine's P.I. was 9, you would score (9/12) x 100% = 75%. The photo here shows an example of a particularly well-designed and successful student-built launcher.

Exercise 3.2   Accurate golf-ball launcher

Aim: design and build a machine that will project a standard golf-ball accurately onto a paper target taped to the floor, at any distance specified by your instructor on the day, between 2 m and 5 m in front of your machine, at intervals of 200 mm. This means that your machine must be calibrated to be able to achieve any assigned distance within these specifications.

The energy you store in the machine must come from your own muscular effort. A trigger mechanism is essential. You may not just pull a lever back and let it go. You have to store the energy and operate the trigger when instructed to do so.

A sample target will be provided for students to inspect. This will be a piece of paper of A3 size, with concentric circles drawn on it, as for an archery target. The bull will be 80 mm in diameter. Carbon paper is placed over the target to record the strike points made by your golf ball. Three shots are allowed, one at each of three distances allocated at random by the supervisor, within the given range.

Your machine may make use of any materials at your disposal, including springs. Immediately before testing, the machine must fit into a cubic box 500 mm on a side, or else it will be disqualified.

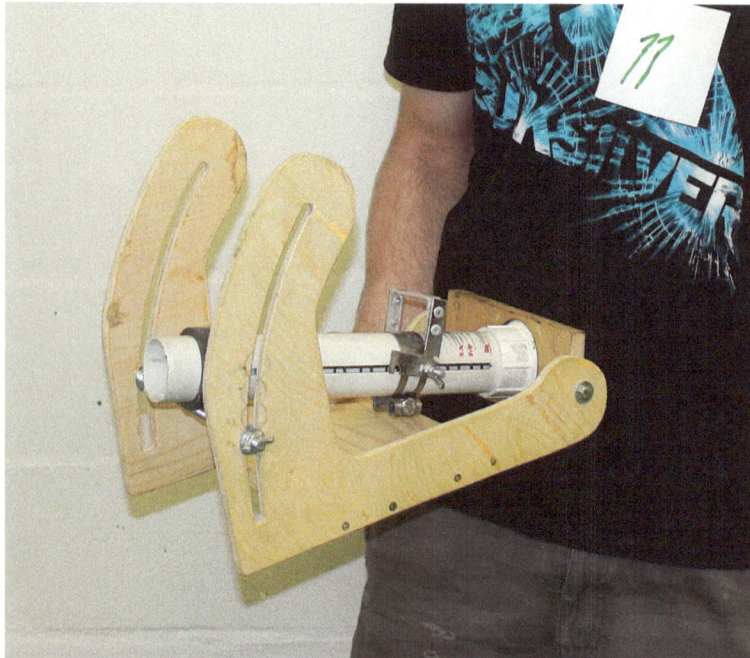

The photo shows one student-built machine with a trigger mechanism, and calibration marks for range.

Exercise 3.3 Vehicle powered by a descending mass-piece

Aim: design and build a wheeled vehicle that will travel as far as possible on a flat, level floor, powered only by a mass-piece of 1 kg descending through a height of 500 mm.

Materials: only the following are permitted: wood, hardboard, glue, string, paper, cardboard, tape and lubricant. Only the mass-piece may be of metal.

The vehicle must have at least three wheels in contact with the floor for the duration of its run, and at least two axles. There is no size or weight constraint, because obviously, the lighter your vehicle is, the better it will perform. The mass-piece you supply will be weighed immediately before your machine's run, so it has to be detachable. If this mass-piece exceeds 1000 g by as much as 1g, you will not be allowed to use it, and will have to use a standard 1 kg mass-piece provided by the department.

The vehicle should follow a straight path without assistance. If it starts veering to one side in its run down the passage, you will not be allowed to correct its direction manually. If it crashes into the side wall of the passage, and comes to a stop, that will be considered to be the end of its run.

Shown here is a particularly efficient machine, which achieved a distance of over 40 m, on an occasion when the next best distance was less than 18 m.

Many different such projects are possible. See Appendix 1 for further suggestions for suitable design-and-build projects. These suggestions are by no means exhaustive. This author has heard of many others. For instance, one colleague set a project for which his students had to build a cardboard boat to travel as far as possible in a swimming pool, powered by a rubber band.

Whatever project one assigns, the instructor needs to ensure that the brief requires students to engage with the specific principles of mechanics that they are supposed to be learning about.

Feedback to instructors after testing day

If desired, one could issue a short questionnaire after the design-and-build project has been completed and assessed, to provide feedback to the instructor, towards improving the briefs for subsequent projects.

Questions like the following could be asked:

- Now that your machine has been tested in operation, and you have compared it to the efforts of others, what do you think you could have done, or ought to have done differently?

- What would you do to improve on your design?
- Is there any aspect of the design brief that you think could have been improved?
- Is there any aspect of the procedure on testing day that could have been improved?
- What have you realised about your own approach to tackling such tasks? Where could you have improved your approach?
- What would you advise the next intake of students to do first when tackling such a project?
- In retrospect, what do you conclude about the process of working with other team members? Was it efficient? Was it effective? What was good about it? What was unsatisfactory about it?
- Now that you have finished the course, of what benefit to you were those design-and-build projects? How much did they help you learn about the principles of mechanics?

The ideal follow-up

If there were sufficient time, it would be ideal to give students a chance to build a second, improved machine, having seen the scope for improvement in their own machine, and having seen what other contestants' machines had achieved. Unfortunately, in my teaching situation, there was never enough time for this. Which is a pity, as in real engineering it is regarded as essential to refine all designs by repeated trials and improvements.

How much should design-and-build projects count?

For any instructor making use of design-and-build projects, there is a pragmatic consideration regarding how much these projects should be made to count in the overall assessment of the students.

If the final mark of their course in mechanics consists of a number of marks for different tests and projects, then, to make sure that students take any given component of the course seriously, that component has to count significantly. The chances are that many students will simply ignore assessment events that count too little, in order to put all their effort into those that count more.

The first assignment that lazy students are inclined to drop will be the design-and-build project, because it demands more of them over a longer period than do other assessed events.

That is why, if you see any value in assigning a design-and-build project, it makes sense for it to count at least 20% of the course mark.

Chapter 4

Using true/false quizzes to stimulate discussion and provide useful feedback

True/false tests do not only have a role in assessment: they can be used very effectively as part of a learning activity.

We'll look at the structure and potential of true/false tests first, and then describe how they can be used.

Obviously, in a true/false test, a statement is made, and the respondent must indicate whether they think the statement is true or false.

At first glance, the scope of this type of question may seem simplistic. It may be reasoned that either you know, or you don't know the answer, so a true/false test might seem to be suitable only for 'knowledge' questions, and not for finding out levels of comprehension or higher states of understanding.

However, a true/false test can be very revealing, depending on the way the statement is phrased and on the scoring method that is used.

Ways of phrasing true/false questions

Here are three commonly used ways of phrasing a question for a test like this:

a. A direct statement of fact, that is either true or false, according to generally accepted knowledge. This type of question is definitely only able to determine something that you may or may not have in your memory. For example:

- James Clerk Maxwell was a professional actor. (true/false)
- The density of copper is approximately 8900 kg/m³. (true/false)

b. A statement that the respondent has to think about, based on other knowledge or judgement he or she possesses, or by calling upon personal experience. This type of question requires some thought, possibly even some synthesis of ideas, e.g.

- The United States is to the east of Portugal. (true/false)
- A person treading water in a swimming pool would be able to push and hold a fully inflated soccer ball entirely below the surface of the water for an extended period. (true/false)

c. A statement whose veracity needs to be checked by calculation or intelligent estimation, e.g.

- If a country's census showed a population of 20 million, and six years later it was found to be 22 500 000, the average population growth would have been approximately 2% per annum. (true/false)
- The value of the velocity ratio of the block and tackle system shown in the diagram would be at least 3. (true/false)

The following table contains a small sample of true/false questions (on randomly mixed topics) that could be used in teaching or testing the subject of mechanics.

1	There are only two states of physical equilibrium.	T	F	?
2	Buoyancy is experienced by objects immersed in a gas.	T	F	?
3	The 'inertia' of an object could be considered to mean the tendency of that object to carry on doing what it is doing.	T	F	?
4	Momentum and inertia are one and the same thing.	T	F	?
5	The value of the friction coefficient between two unlubricated given solids can range from near zero to just less than 1.	T	F	?
6	Newton's first law is about equilibrium.	T	F	?
7	The Law of Conservation of Momentum contradicts the Law of Conservation of Energy.	T	F	?
8	Newton's Laws were known to Galileo.	T	F	?

The '?' option denotes 'I don't know'. The reason for including this option is explained further on.

To use a quiz like this as a learning exercise in class time, a paper copy of the quiz is issued to each subgroup of three or four students. They have to discuss how they would answer it, and submit their best collective effort after a given time, with the participants' names indicated clearly in a demarcated place on the quiz paper. Choices are to be indicated in ink, with no erasure permitted.

Papers are then swapped between groups or assigned by the instructor to different groups, for marking. If students mark the quizzes, their names should be clearly indicated on the quiz paper that they marked.

Answers are provided by the instructor. Once each quiz is marked, it is returned to its originating group.

Each group is then given a chance to ask the instructor one question arising from the quiz, or arising from the way their paper was marked, in front of the whole assembly. This allows the instructor to clear up those misunderstandings that clearly have survived the attempts by group members to arrive at the correct answers. Which, in turn, means that the instructor doesn't have to explain everything about the topic in hand, and needs to address only those questions whose answers the students were unable to provide by collective effort in their groups.

The procedure that has been described above is clearly a form of tutorial, with the quiz as the stimulus material.

The value of sub-group discussions that attempt to answer true/false questions should not be underestimated. By arguing and explaining, students learn from one another.

The result of an exercise, when used as described above, does not necessarily need to contribute to the assessment of the students. However, if desired, it could.

The type of scoring system to use with a T/F test, if it is used for assessment:

The method that doesn't work

A scoring system that is commonly used with true/false questions awards one mark for a correct answer and zero for an incorrect one. That method is completely unreliable, because a candidate who did not know the answer to any of the questions in the quiz could, statistically, get 50% for the test simply by random guessing. If this happened, it would not represent in any way the true state of that student's (or a group's) knowledge.

The method that does work

The following scoring method, which I have found to be very effective, *does* probe for the true state of the candidate's knowledge, and most importantly, encourages students to develop honesty about what they do or don't know. It works like this:

Each correct answer is rewarded with 2 marks.
If you say you don't know the answer, your honesty is rewarded by being allocated one mark.
If you omit to answer, nothing can be deduced about your state of knowledge. You might not understand the question, or you might understand it but have no clue about how to answer it. Maybe you ran out of time and didn't get to that question. Or, you might be leaving it until later to come back to it and have another think. Either way, for omitting to answer, you are awarded zero.
Incorrect answers are penalised, to discourage guessing. Why? Because in engineering, it is unacceptable to guess. A wrong guess on the job could result in damage to property, failure of design, waste of resources and danger to personnel. For each answer that you get wrong, 3 marks are deducted.

If you introduce this marking scheme without an explanation of its purposes, it will at first be very unpopular with students, because they imagine you are hell-bent on failing them.

However, if you issue a clear explanation of your motives in writing, as well as explaining these motives in person, you will see an interesting development as the term progresses. After a few tests marked in this way, students are far less inclined to guess, and more motivated to take responsibility for their own learning. In my experience, when exposed to this scoring method, students learn very quickly to be more honest about what they know.

It begins to come home to them that learning engineering is not served so much by trying any means possible to produce the answers desired by their examiners, as by building a personal knowledge structure that is consistent.

If the testing process leads students to developing a greater level of integrity about their own understanding of the subject, this definitely helps to develop their capacity to become responsible engineers.

An additional advantage of this scoring method is that it enables the instructor to see directly from the test results which questions the student body as a whole admits to being unable to answer. Then the instructor can do something remedial to address the areas of confusion.

The implications of using the 2,1,0, −3 scoring system:

Say, for instance, that you, as a candidate, have to tackle a test with 100 questions, and cannot answer any of them with confidence.

We have already mentioned that, if you took a wild guess at each question: applying the commonly used scoring system (right = 1 and wrong = 0), if you got lucky half the time, by sheer chance, you could score around 50%, and might even score over 50% and so pass the test.

Applying the 2,1,0, −3 scoring system, if you guessed all the answers and got lucky half the time, you would score 50 times 2 for your (luckily) correct answers, plus 50 times (−3) for the ones you got wrong, namely a total score of −50.

However, if you were honest and admitted you did not know any of the answers, your score would be +100. So, any score over 100 would be a broad indication that the examinee had been tolerably honest about his or her state of knowledge. This is in itself a recommendation, because it provides a dimension of constructive feedback to the student.

You could think of it conversely: provided that all the quiz questions had been answered, any score below 100 is likely to have been achieved by some measure of guesswork.

Supposing you knew your work and got all the questions correct. In that case, you would score 200. If you knew half of the answers and admitted to not knowing the other half, you would score 50 times 2 for the ones you got right, plus 50 times 1 for the ones you did not know, namely a total score of 150.

A pass-mark for such a test might therefore be chosen (arbitrarily, as pass-marks always are) to be somewhere between 150 and 200 marks.

Clearly, when making use of this scoring system, a mind-set has to change, in order not to think that marks need to be on a linear scale.

True/false questions scored in this way are actually ideally suited to testing students' knowledge of the *principles* of mechanics. One might think there is a limited range of questions that can be posed, on account of the relatively few principles that govern the science of mechanics. However, with some creativity, one can generate enough variety to keep students on their toes. This author's book 'Revision Exercises in Basic Engineering Mechanics' *(see Appendix 5)* contains forty-five true/false quizzes, each containing ten statements, grouped according to topic. These quizzes may be used without requiring permissions.

An alternative way of arranging the questions in a T/F test

A variation on the structure of a T/F test is the following: an incomplete sentence begins the statement, which is followed by several offshoot possibilities, each forming a complete statement when combined with the opening phrase.

The candidate might be marked on each offshoot statement individually, or might need to get the entire set correct in order to be considered competent on that particular topic. Three examples follow:

1. An object is in equilibrium if it is...

a	Standing still relative to the Earth.	T	F	?
b	Moving with constant angular velocity in a circular path.	T	F	?
c	Floating on still water without moving.	T	F	?
d	Sinking in still water at slow constant velocity.	T	F	?
e	Accelerating uniformly on a straight path.	T	F	?
f	Spinning on an axis that passes through its centre of gravity, with constant angular velocity.	T	F	?
g	Moving in a straight line with constant velocity.	T	F	?
h	Falling in still air, having reached terminal velocity.	T	F	?

2. If a set of forces is NOT in equilibrium, its resultant...

a	Must be greater than any of the forces in the set.	T	F	?
b	Must be zero.	T	F	?
c	Could turn out to have a value smaller than any of the forces in the set.	T	F	?
d	Can be determined by drawing a polygon of forces.	T	F	?
e	Can be determined by a summation of components.	T	F	?
f	Always has the same magnitude as the equilibrant to the set of forces.	T	F	?

3. A Free-Body Diagram...

a	Is a diagram showing all the forces acting on an object.	T	F	?
b	Can have any number of forces shown on it.	T	F	?
c	Shows all other objects which are touching the central object.	T	F	?
d	Shows only the forces exerted by the object.	T	F	?
e	Must have all the forces in it reduced to their rectangular components.	T	F	?
f	Can only be drawn for an object which is in equilibrium.	T	F	?
g	Shows all the forces acting on the object, as well as all the forces exerted by the object.	T	F	?
h	Should not include the weight of the object.	T	F	?

Here follow two examples of true/false tests on single topics. These are extracts from this author's book 'Revision Exercises in Basic Engineering Mechanics':

	True/false Test # 5a Centres of mass, centres of gravity and centroids			
1	It is possible for the centre of gravity of an object to lie outside the material from which the object is made.	T	F	?
2	The centre of gravity of an object can be situated in a different position from that of its centre of mass.	T	F	?
3	Any calculations to determine the location of the centre of mass of an object are based on a procedure for determining the location of its centre of gravity.	T	F	?
4	The location of the centre of gravity of a rigid body does not change, unless the body is deformed or has material added or removed.	T	F	?
5	When determining the location of the centre of gravity of a complex-shaped flat plate of variable density, plate areas can be used in place of masses.	T	F	?
6	If some material is removed from an object, it is impossible to determine how the object's centre of gravity has shifted.	T	F	?

7	The centroid of a volume could lie at a different position than the centre of gravity of a solid object with the same dimensions.	T	F	?
8	To locate the centre of gravity of a structure made of straight rods, such as a truss, one first needs to know the location of the centres of gravity of the constituent rods.	T	F	?
9	It is not possible to determine the location of the centre of gravity of irregularly-shaped objects.	T	F	?
10	You can determine the mass of a long object such as a heavy beam, if you have only one weighing scale that can weigh up to just over half the weight of the beam.	T	F	?

	True/false Test # 10a **Motion influenced by gravity**			
1	The reason that a feather dropped from a given height takes longer to reach the ground than does a small stone with the same weight as the feather, is that the respective air resistances on the two objects differ.	T	F	?
2	The value of the gravitational acceleration, g, is always 9.81 m/s^2, at all locations on Earth.	T	F	?
3	For a falling object, the terminal velocity is that velocity at which the air resistance to its motion is equal to the weight of the object.	T	F	?
4	The height of a cliff can be estimated by dropping a stone from the top of the cliff and measuring the time it takes to reach the ground below. This value of time can be used in one of the equations of motion to determine the displacement of the stone from the launching point.	T	F	?
5	The trajectory of a projectile can be plotted if the launch speed and launch angle are known.	T	F	?
6	Air resistance plays a minimal role in affecting the trajectory of a high speed projectile.	T	F	?

7	Ignoring air resistance, a projectile fired at a given speed over horizontal ground will have the same range when the launch angle is 23° as when it is 67°.	T	F	?
8	The effect of air resistance on a projectile is always more noticeable in the vertical direction than in the horizontal direction.	T	F	?
9	It is impossible to determine the launch speed of a projectile such as an arrow, by purely mechanical means.	T	F	?
10	If James and Peter both throw a cricket ball at the optimum trajectory, and James can achieve a throw distance of 80 m, while Peter can achieve only 60 m, then the difference in their launch speeds must be 20 m/s.	T	F	?

To sum up this chapter:

If marked in the 2,1,0,−3 pattern described in this chapter, true/false quizzes can be used, not only for assessment, but:

- To probe for students' grasp of issues that go beyond simple knowledge, to comprehension and synthesis,

- As stimulus material for tutorials,

- To provide feedback to students about what they don't understand,

- To provide direct information to instructors about which issues the class as a whole might need clarifying, and

- To train students to become more honest with themselves about their own understanding of a topic.

Chapter 5

Using previously unseen calculation questions as the basis for tutorials

In many, if not most engineering schools, calculation questions usually comprise the bulk of what students will face in the examinations. To help prepare students for the exams, such questions are frequently used as the focus of tutorials.

However, as described in Chapter 1, the way in which such questions are employed in a tutorial makes a great deal of difference to the effectiveness of the event as a learning activity.

One of the essential features of a productive tutorial is the introduction of previously *unseen* stimulus material. The second essential feature is that a specific task needs to be set up for the students to engage with that stimulus material.

Calculation questions can definitely be used as stimulus material. Having to solve 'new' calculation exercises in a collaborative discussion group of three members gives students a chance to become aware of and to practise the type of thinking required in order to work towards a solution.

The ultimate aim, naturally, is to enable students to become competent to solve such exercises on their own. However, collaboration within sub-groups provides students with confidence about the approach they need to take when attempting to solve similar exercises individually.

Each institution and each individual instructor will have their own style of calculation exercise, and their own expectations of how these exercises should be tackled.

A small sample of calculation exercises devised by this author is provided at the end of this chapter. From these can be seen the importance that this author gives to providing illustrations that define the practical circumstances of the problem with clarity.

For unrestricted use by instructors, over 500 original illustrated calculation exercises, with answers, can be obtained from my other books on mechanics that are described in Appendix 5.

The purpose and scope of a calculation-type exercise

The *purpose* is to be able to apply scientific and mathematical reasoning in a quantitative way to engineering problems.

The *scope* is that realm of problems of a kind that are amenable to being addressed using mathematics. Not all engineering problems are.

It is important to be clear about the role of mathematics when applied to aspects of reality. Pure mathematics describes a world in which neat, orderly, predictable and consistent relationships exist between variables. When you obtain an answer to an exercise in pure mathematics, if your process has been correct, that answer has to be trusted. It is exact.

However, when mathematics is applied to engineering situations, a certain amount of wisdom has to be invoked in order to deal with the practical limitations of the result of the calculations.

These limitations occur because calculations mostly do not culminate in a result that is the complete solution to a practical problem. The results of calculations only provide *a guide* to what needs to be decided next. This is true whether the patterns that describe how the variables are related are derived from theorising or from empirical observations.

A mathematical formula very rarely represents the full reality of an situation.

The laws of mechanics, and their derived conclusions, often result in equations that students can come to regard as 'formulas' into which figures can be plugged, from which answers emerge, that they can write down in false confidence, feeling that they have exercised their full responsibility.

However, formulas are not magical incantations that lead to 'engineering' having been done.

People in engineering practice have a a mildly amused skeptical attitude to fresh graduates arriving, armed with formidable 'cure-alls' in the form of the equations they have learnt to use.

There are several ways in which students need to learn the limitations of formulas.

1. Having to make too many assumptions can lead to diminished confidence in an answer.

In order to make use of a formula, certain assumptions may need to be made. Learning to make reasonable assumptions is one of the aims of an engineering education.

Typical assumptions that sometimes have to be made before a mathematical solution is applied to a problem could include, for example, that:

- Friction is minimal,
- Temperature changes do not affect the situation,
- The materials in question are rigid,
- Certain forces can be assumed to act at a point,
- The spring does not reach its elastic limit,
- This short curved path approximates to a straight line,
- A value obtained from reference sources is accurate, and
- The two-dimensional structure shown is stable in the plane of the paper.

One can learn something about which assumptions are reasonable from the way in which an instructor makes assumptions when demonstrating his or her approach to a calculation problem.

However, students need to become aware that some assumptions can lead to a potentially wide deviation between the calculated answer and what could be expected to occur in practice. This kind of deviation is more likely to be large when several assumptions have been made in order to tackle a given problem.

The most extreme example that I can recall occurred in a class in thermodynamics that I attended as a final-year student. The lecturer demonstrated his analytical method of solving a particular problem. In this problem, water at a certain temperature was made to flow down an angled flat plate in ambient air that was at a different temperature to that of the water on entering. The objective was to determine the water temperature by the time it reached the bottom of the plate. The man went through about six assumptions and forty minutes of calculations, using the most complex mathematics, involving nearly every letter in the Greek alphabet. I lost his train of thought after thirty minutes of hard concentration. When he finally arrived at his answer, I put up my hand and asked him how accurately the result of this calculation might reflect what would be found in practice. His answer: 'Within, plus or minus forty percent'. I was astounded, and said that, instead of going through this whole lengthy and complicated process, it would have been quicker and more reliable to set up the apparatus and *measure* the resulting temperature. The lecturer looked incredulous, as if this ignoramus in front of him hadn't appreciated the beauty of the theorising to which he had just treated us. True: I hadn't.

2. Impracticality of precisely calculated values

A second reason for caution about a calculated result is that the degree of precision implied by the calculated result may be impractical. If you, say, determine the value of a force, in newtons, that is needed in a given situation, accurately to eight decimal places, when the accuracy with which you can create that force is to the nearest newton, then the eight decimal places mean nothing.

3. Factors of safety

A third reason for caution about a numerical result is that the practicality of an engineering situation requires a sensible factor of safety to be used. You might have obtained a answer to a calculation telling you that the diameter of a bolt needs to be 7.3256941 mm, but you need to make a choice among the sizes of bolt that are actually available, and you may need to estimate a suitable factor of safety. You might therefore select a 10 mm bolt for your solution to the given problem. In which case, the calculated answer was just the first step towards suggesting a practical solution.

For these reasons, it is a good idea to add sub-questions wherever possible to calculation questions that are used in tutorials. The sub-questions could be designed to get participants thinking about the estimates they need to make in order to render their solutions practically plausible.

The same applies to aspects of lab experiments. For example, one typical first-semester experiment requires students to determine the coefficient of friction between two given surfaces. They will happily show from the gradient of their graphs that the value of this coefficient, is, for example, 0.45785. However, if they repeat the experiment immediately afterwards, using the same apparatus, this value might come out with only the first, or, at best, the first two places after the decimal point being in agreement with the original value obtained. Students should be prodded about the degree of accuracy that is actually obtainable for such a coefficient.

For engineering students, it is sound practice to see all calculated results in terms of how they relate to plausible reality, not just to theory.

A useful procedure, when making use of previously unseen exercises in tutorials

1. Allocate students to sub-groups of three.
2. Issue the instructions for what needs to be done in the session.
3. Issue the questions to be solved.
4. Circulate to answer queries while the sub-groups tackle the questions.
5. Require each sub-group to submit, in writing, their answers (not their workings) to all of the questions, together with an estimate of the level of confidence they have in their answers.
6. Allow all present to see all the submitted answers.
7. Allow a few mintues for each subgroup to discuss the displayed answers.
8. Indicate your own answers to the questions.
9. Invite one question from each sub-group to be put to the tutor and aired to the assembly.
10. Announce the details of a test on this work that will be administered within the next few days.

The use of calculation exercises in assessments

The traditional examination question used in engineering courses requires a candidate to recognise a standard 'type' of problem, and perform calculations to obtain a 'correct' answer. Credit is also sometimes given for knowing formulas, the accuracy of calculations, or demonstrating an appreciation of the meaning of the answer.

The use of calculation questions to assess students has definite attractions, but it also has pitfalls, and both need to be acknowledged.

We will list the attractions first, and then present the cautions.

Good reasons to make *some* use of calculation questions

1. During a test, calculation questions allow candidates to recognise some of the problem-types they have encountered in class. This creates an opportunity for candidates to show they can do what has been practised, which builds confidence at the time of the test.
2. The candidates' workings and answers provide examiners with insight into the general calibre of each student's mathematical reasoning abilities, which need to be of the standard expected in the institution.
3. Calculation questions are easy to mark, provided that the candidates know their work and can set out their reasoning logically and neatly enough to follow.

4. If accompanied by a suitable illustration, a calculation-type question can provide relevance to something that might conceivably be encountered in real engineering. It wouldn't be able to take everything into account that contributes to the reality of an engineering situation, but it goes some way towards doing so. In this respect, it makes candidates feel that there is a useful application for the principles they have learnt, and that they are not just dealing with abstractions.
5. If a numerical answer has to be obtained, that answer enables test-takers to compare their respective answers immediately after the assessment event, to discover from peer feedback whether or not they have been on the right track.

Concerns about the use of calculation-type questions in assessments:

Concern #1: The delusion of objectivity

To an examiner, the main advantage of using calculation-type questions is that, assisted by your moderator, you can establish what the 'right' answer is for each question, so you set out marking with some measure of confidence that you have an 'objective' criterion against which to measure the candidate's performance.

However, your confidence in such 'objectivity' may lose some shine when you realise that both the nature of the question *and* the definition of an acceptable answer came into being after a number of *subjective* choices, namely:

- The choice of what question to ask (millions of possibilities exist),
- How closely the question resembles other questions which the candidates may have encountered (is it new or practised?),
- What level of difficulty in the question you deem appropriate,
- The way you phrase the question,
- What information you provide,
- Whether or not you provide an illustration to illuminate the circumstances that prevail,
- How informative and clear that illustration is,
- What steps you expect the candidate to follow in solving the problem,
- What level of accuracy you will tolerate,
- What actions you will award marks for,
- What you will deduct marks for,
- Which qualities in an answer you consider to be evidence of competence,
- How much time you allocate, and
- What resources you allow the candidates to have at their disposal during the test.

The decisions that examiners make about all of the above items are inevitably, *always* subjective. When designing calculation-type questions, you may take as a guide the style of the traditional questions used in your department, or rely on your own professional judgement, but you have to acknowledge that the decisions such as those listed above are all decisions that *you* make. Any other examiner might employ a significantly different set of criteria in making such decisions. Hence, true objectivity is not really attainable.

Consider this extreme example of when objectivity went out of the window in a calculation-type subject: A friend told me that his mathematics teacher in high school declared he would never award a student 100% for a maths test. No matter how simple the test, or how well you answered the questions, even if you got them all correct, he refused to award more than 99%. His 'reasoning' was that no-one could know all there was to know about mathematics. That was mind-boggling illogic, but he had the power to pass or fail, and could not be argued with. If you are an examiner, you have the same amount of power.

If objectivity is not attainable, that doesn't mean that examiners should avoid setting calculation-type questions. They just need to keep in mind (a) that they are fully responsible for all the decisions listed above, and (b) that a solution is not automatically 'objective' simply because calculations were involved.

Concern #2: Can the answer to a calculation-type question be relied on to reveal a candidate's thought process?

A particular benefit enjoyed by markers is that they can save time by flicking through the candidate's answers and awarding full marks to all those questions that have the correct answer, without needing to check whether the logic of every step in the solution was fully explained.

While having such an opportunity is a great convenience for the marker, it is not an infallible assessment of the thought process employed by the student. If the answer is correct, there is certainly a strong possibility that the student knew what he or she was doing, but that is not a certainty, because students could produce correct answers as a result of:

- Having worked through the very same question the day before the test, and having memorised the steps, while unable to apply the appropriate logic to other questions of a similar nature, or

- Cheating by copying another's work or by bringing crib-notes to the venue, or by having an electronic communication connection to someone outside the venue.

- Although rarely occurring, I have encountered a few times in my teaching career, instances in which a 'correct' answer was obtained by the candidate making two mistakes which cancelled one another out. For example, misreading a calculator readout, and thus inadvertently multiplying by a thousand somehere in the calculation and then, later on, mistakenly dividing by a thousand.

So, a 'correct' answer given by a candidate is not necessarily indicative of the reasoning that led to that answer being correct. This issue makes it essential for a person marking the work to cast an eye over the reasoning the student has employed, and not just accept the 'correct' final answer as representing competence in the knowledge being tested.

Concern #3: Could arithmetic and mathematical errors have masked the fact that the candidate actually understands the principles of mechanics?

On a course in mechanics, is the point of any exercise to get the mathematics right, or to get the principles of mechanics right?

Students who do not get the acceptedly 'correct' answer may very well have known which principles to apply and what method to use, but may have made mistakes in their calculations. It is easy to make an arithmetic mistake due to being distracted by hastiness or tension experienced during exam conditions.

A student could also make a mistake due to incomplete understanding of the mathematical processes he or she is trying to use, like forgetting the formula for the integral of a certain function. Neither of these two causes have anything to do with the student's grasp of the principles of mechanics.

Concern #4: Are the imposed time-constraints realistic?

It is common examination practice to expect that students should be able to get all their calculations correct in the available time, in order to be considered competent. Imposing such an expectation is a necessary feature of the logistics of conducting an exam. However, it does not really match the conditions which will apply in the conduct of engineering. No company would risk relying on a one-off calculation done at high speed and under duress, by one person. All significantly important calculations are checked, discussed, re-done and checked again.

Concern #5: What do you allocate marks for, and would other examiners agree?

The way that marks get allocated for answering a calculation-type question can vary greatly. Some examiners allocate marks for certain steps that they expect to see carried out in the solution. These 'steps' can be detailed or broad.

I knew one examiner who went so far as to allocate one quarter of a mark for each of the various stages of a solution, thus giving himself a great deal of intensive work. In a question worth ten marks, the marker would have to make 40 separate decisions. There is such a thing as creating too much work for yourself.

A more effective way of assessing the answer to a calculation question is not to award marks for detailed steps in a calculation, but to award a mark for your *impression* of the way the candidate has fulfilled each of the following three broad criteria. For a question worth ten marks:

Did the candidate take and describe a valid approach? (allocate 0 to 4)
Did the candidate carry out that approach properly? (0 to 4)
Did the candidate provide a credible answer? (0 to 2), totalling 10 marks.

If you suspect that doing it this way is allowing too much room for the assessor to allow his or her subjective (but professional) opinion to override 'factual evidence', consider the outcome of the following trial:

This author once marked a batch of 36 papers in Mechanics 1 in the way described above, without writing anything on the papers. A colleague then marked the same batch, mark by mark for each step taken, according to his detailed memorandum, and the scores were later compared. Not one candidate's score for the test differed by more than 3% between the mark I allocated and the one he did.

Concern #6: Does having to answer calculation-type questions reflect the reality of what the candidates will end up having to do in their jobs as engineers?

The widespread use of such questions in the assessments used by engineering departments gives students the impression that an engineer's work consists of solving clearly defined problems, in particular ways, using a set of appropriate data that has been provided.

In real engineering practice, decisions that need to be made rely on optimising one's way through a variety of issues that need to be accommodated. Seldom is a decision based on one narrowly-defined calculation. Certainly, calculations are often needed, but the results of calculations don't solve entire problems, they merely provide clues about how to optimise some of the issues which need to be taken into account.

Concern # 7: Some calculation questions that students face will inevitably, *intentionally* depart from the full reality of an engineering situation.

Here is an example from basic vehicle dynamics: Data is supplied about a car maintaining a steady velocity up a specified incline. Candidates are tasked with determining the values of such variables as the speed of the car and the power and torque developed by the engine at this speed. At a basic level of mechanics, test-takers are expected to do this without taking into account the torque curve or the equation relating air resistance to the properties of the vehicle shape and size.

These important factors are omitted for the sake of simplifying the calculations, so that the student can demonstrate knowledge of how to apply certain principles (in this case, how to apply the equilibrium of forces in the direction of motion, and how to determine the driving force at the wheel perimeter as a function of the engine torque, the overall transmission ratio and efficiency).

Since some of the considerations that actually have a significant influence on the authenticity of an answer have to be brushed aside or assumed inoperative, the way many questions are phrased doesn't really reflect the full reality of an engineering situation.

To sum up this chapter

If you thought that the use of calculation-type questions was entirely sufficient to help you identify which candidates deserved an engineering degree, it may be time to adjust that opinion.

Purely on account of the strength of tradition, it is unlikely that the use of calculation-type questions in assessments will ever be abandoned. However, they can very justifiably be supplemented by the use of some of the other assessment exercises described in this book.

However, calculation-type questions can definitely be used in a developmental sense, when employed in tutorials, where they can provide the stimulus as well as the aim for the discussion.

Some Sample Calculation Exercises

Here follows a sample of three calculation exercises. The way these are presented reflects the author's belief in creating exercises that have relevance to something practical, and in illustrating them clearly.

The first two are 'standard' exercises, in which all the necessary data is given. The third one is different, in that not all the data needed to solve the problem is provided. This one is designed to be solved by discussion in a small group. It requires some assumptions to be made, and intelligent estimation to be applied.

Ten further examples are provided in Appendix 2.

Exercise 5.1

A two-metre-long wooden pole, of density 800 kg/m³ and diameter 160 mm, rests on a level floor.

One end of the pole is raised by an attached cord that passes over a sheave and supports a mass-piece of weight W.

The pole reaches equilibrium in the position shown. Assume the friction in the sheave is neglible.

Determine the weight of the pole [315.6 N], and the mass of the hanging mass-piece. [19.19 kg]

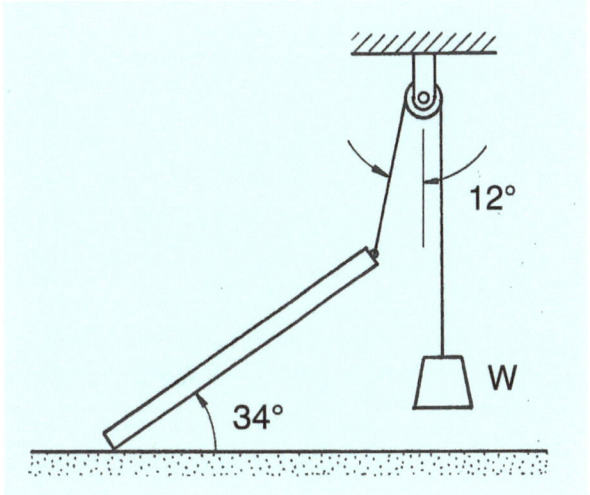

Exercise 5.2

A two-dimensional frame is made of steel bars welded together at their ends.

The table shows the weight per running metre of the materials from which each respective bar is made.

Determine:

- The lengths of all the bars, to the nearest mm, and their weights.

- The location of the centre of gravity of this frame relative to the x- and y-axis, to the nearest mm.

[X = 346 mm; Y = 801 mm]

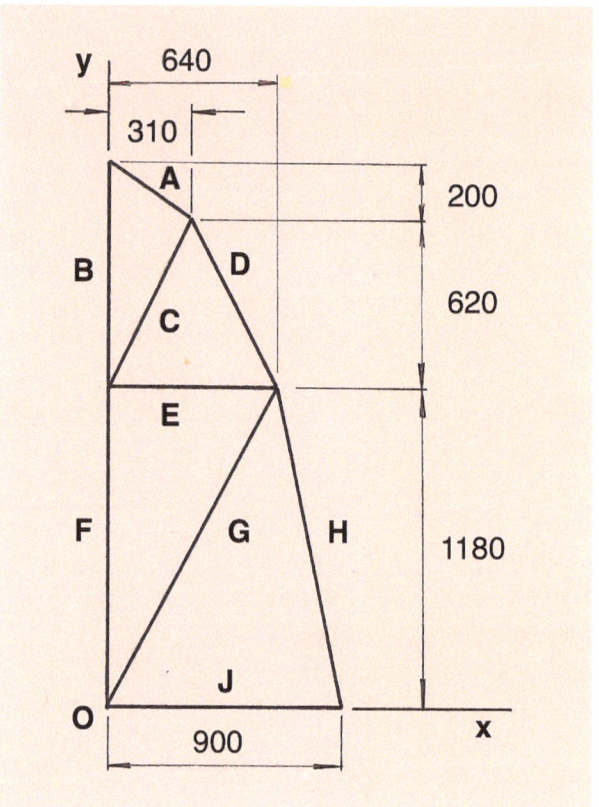

Bar	length [mm]	Weight of material [N/m]	Weight of bar [N]
A		30	
B	820	30	
C		30	
D		30	
E	640	40	
F	1180	40	
G		40	
H		50	
J	900	50	

Exercise 5.3 (answers not supplied, as this one is a group exercise)

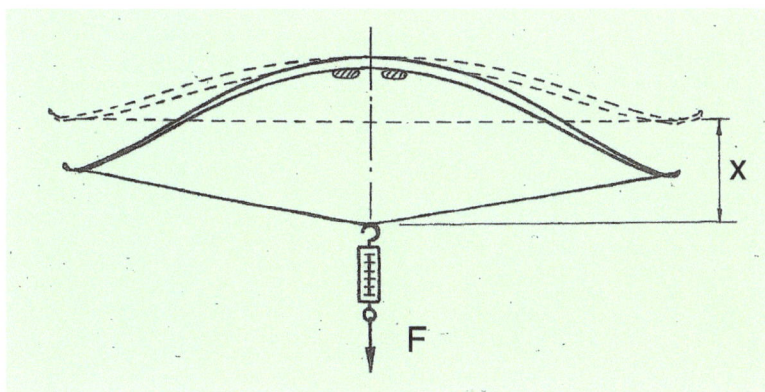

An archery bow is extended by pulling down on the string such that the force **F** required to draw it down, and the corresponding extension **x**, can be measured.

For this bow, the maximum draw force is 150 N when x = 600 mm.
The bow-string is 1.8 m long.

a. Estimate the amount of energy put into flexing the bow at full draw.

b. Draw a graph showing the force that the bowstring exerts on the arrow from the moment of the string being released until the moment the string is again taut.

c. Estimate the launching velocity of an arrow of mass 60 grams if the transfer of energy is assumed to be 80% efficient.

d. Estimate the maximum range of this arrow over level ground.

Chapter 6

Supervising experimentation in the mechanics lab

Participating in lab work can be very beneficial to students. However, the effectiveness of lab work and the motivation to engage in it depends on the way students are expected to use their minds during lab sessions.

Unfortunately, there are several reasons why many instructors have little appetite for conducting lab work with large classes of first- and second- semester students. The consequence of less-than-mild enthusiasm from instructors is that students tend not to take lab experimentation seriously.

Problems include:

- Issues of logistics: again, too many students, too few resources, especially a shortage of competent lab assistants,
- Repetitiveness of the prescribed experiments for those supervising labs. If you think you can do them in your sleep, you probably are!
- The procedure during labs is found boring by students, because they are not expected to use their initiative, and usually just have to carry out a set of instructions,
- Tedium for those who mark lab reports, due to repeatedly encountering the same kinds of glaring mistakes and misconceptions, particularly knowing that such mistakes have been cautioned against in the lab instructions, and
- If lab work doesn't count much in the final assessment mix, students will not take this work seriously.

There is little chance of experimentation being found meaningful by students or instructors in such circumstances.

The actual activity of doing an experiment needs to be challenging and interesting, or else lab work in an institution will inevitably spiral into disuse.

'Disuse' is no exaggeration. When I started lecturing in 1976, I was asked to revive the mechanics lab, which apparently had not been used for some time. No-one could tell me how long ago it had last been used. While exploring the store-room attached to that lab, I found some brand new apparatus, still in its packaging, together with the catalogue from the manufacturer, dated 1924! Asking around, I discovered that my colleagues regarded having to organise lab work as a nuisance. They could prepare students for the national examinations without doing lab work, so they avoided any mention of labs.

I had to begin working with some very dated lab instructions, and try to design the activities so that students saw value in doing experiments. Recalling my own frustrations with lab work when I was a student, I set out to see if I could improve my students' experiences in the lab.

I noticed that the kind of lab activities that are problematic are sessions in which the procedure and equipment are fully prescribed, and the expected result is virtually demanded.

In sessions like this, students are easily tempted to 'cook the books' by adjusting the findings to suit the expected result. The way they use their brains is minimally beneficial, and the temptation to take short-cuts is always present.

At the opposite end of the scale of responsible involvement would be to carry out a completely original experiment, such as is usually expected from masters students. Naturally, one does not start beginners on completely original experiments, because they would not be up to the task without practice in various skills, such as:

- Understanding the importance of stating the aim of an experiment clearly
- Making reliable measurements of basic variables with the equipment at hand,
- Understanding the potential sources of error when making measurements,
- Plotting graphs that illustrate relevant trends in the data,
- Interpreting data, and
- Making appropriate conclusions that refer back to the stated aim.

Getting practice in these skills is a necessity, which is why certain standard experiments need to be done before students can embark on original experimentation.

Let's assume, therefore, that in a lab programme of a mechanics course, a number of experiments will be expected to be performed. The issue is then:

How to make experimentation interesting and meaningful to students

There are several strategies that help to accomplish this:

Strategy 1: Provide good quality supervision.

Ideally, it is advisable at all times for a knowledgeable and involved person to supervise every group doing an experiment. This person, whether instructor, tutor or lab technician, should understand how to give appropriate feedback, and be capable of asking the group of experimenters leading questions, in order to get them to realise for themselves why things are done the way that they are.

The supervisor needs to keep a close eye on proceedings, to give just the right amount of guidance, and to challenge the group to defend the reliability of their measurements and the inferences they are making from the data. If they have been properly guided, students will understand what is important, so they won't write lab reports with glaring errors.

However, sufficient numbers of competent, interested and involved supervisors cannot always be provided. When I was teaching, I supervised every lab session myself. Sometimes I had a senior student as an assistant, whose help was very welcome, but I was still in the lab to guide events, by providing a presence of authority that lab assistants couldn't.

If my classes were too big for all the students to be in the lab at once, I split the classes and had to hold more lab sessions.

Strategy 2: Issue clear instructions.

To avoid having students milling around, uncertain of what to do when they enter the lab, it is essential to have a clear set of lab instructions issued at the start of term.

My lab instruction booklets described how the lab system worked, and provided brief instructions for all experiments. The contents of these booklets evolved from semester to semester, as my lab system became more streamlined.

In some circumstances it may be found useful to include a paragraph like the following:

NB: Read the lab instructions for each experiment before coming to the lab to perform it. The lecturer won't explain the method. He will answer questions and offer help and advice, but cannot spare the time to re-explain what you have received in print. If you have to spend time finding out what to do, you may find yourself unable to complete the experiment in time. The lab will close at 16:30 whether or not you have finished.

At the start of the lab session, I would asking the class leading questions about the experiment, before the work commenced. For example:

What is this experiment trying to accomplish?
Can you describe the aim precisely?
What's the general procedure you have to take?
Any questions before we start?

Strategy 3: Provide instructions that leave room for some initiative.

Here follows an example of a set of instructions for one experiment. Step 'a' in this set of instructions shows clearly how part of a basic experiment can be left to the students' initiative.

Experiment 1: The block and tackle

a. Set up an arrangement of tackle similar to the one illustrated here. For this purpose you may use one of several sets of blocks, each containing one, two, or three sheaves. Work out how to thread the tackle and how to suspend the arrangement without it toppling or getting tangled.
b. Apply a load, L, and raise this load through a measured height, h.

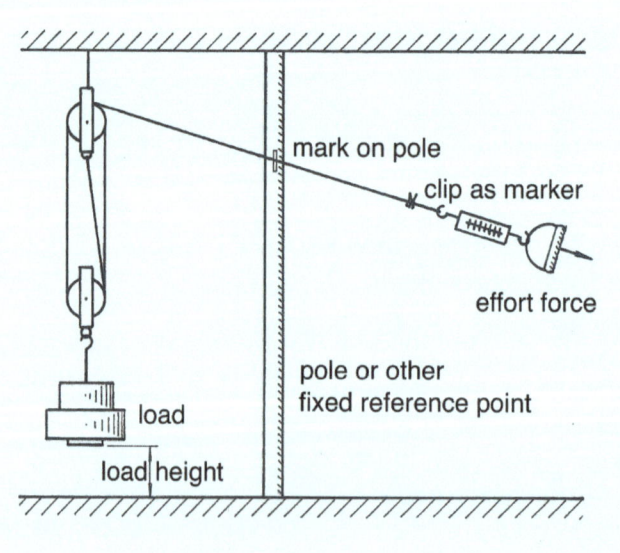

c. Record the value of the effort force, E, using the in-line spring scale, and record the distance moved by the effort rope in order to raise the load through height h. This distance can be measured by attaching a small clip to the rope, level with a fixed point at the start of the pull, and attaching another clip level with that fixed point, at the end of the pull, then measuring the length of rope between the clips.
d. Repeat the procedure for five different loads, increasing the load each time.
e. Plot a graph of effort vs. load, to scale.
f. Determine the 'Law' of this machine.
g. Plot a graph of efficiency vs load for this lifting machine.
h. For the particular operation of the machine when the largest of all the loads was raised, draw up an energy accounting diagram, to scale.
i. Make a note of all the potential inaccuracies you can detect in the way any of the variables are measured.

Strategy 4: Suggest that the students expand on the aims of a basic experiment.

One experiment that is very common to courses in basic engineering mechanics is the one which investigates the relation between the limiting friction force and the normal compressive force, namely $F_{max} = \mu N$.

As soon as the basic experiment has been performed, one can ask the class how it could be expanded. Could they do the same with different pairs of materials? If there is time, one could provide them with blocks and planes that have a suitable variety of facings. Perhaps suggest that they try to find out whether the values they obtained for the coefficient of friction agree with published values, and if they agree with the values obtained by other groups. Any discrepancies found between the results obtained by different groups should lead to students taking a closer look at their own experimental method.

Strategy 5: Encourage students to show initiative in experimental design.

For instance: in connection with investigating the relation of variables around sliding friction, students often are puzzled by the assertion that the amount of contact area between the two sliding surfaces does not affect the value of the friction force. One might ask them to design a variation of the above experiment in which that assertion is tested. This will usually involve them suggesting modifications to the apparatus. Some students enjoy the opportunity to make creative design suggestions, and others enjoy criticising their suggestions. This gives rise to lively discussions. Ideally, it would also give rise to a plausible design being made and tested.

Strategy 6: Provide an innovative piece of apparatus to investigate a phenomenon for which the outcome can't be predicted.

With some creativity, one can provide lab experiences that are interesting and enjoyable. For instance, I commisioned our chief technician to build a piece of equipment to investigate the shape of the trajectory of a golf ball projected with a small velocity over a short range. Does the trajectory conform to a parabolic shape, as suggested by the theory of projectiles? The excuse is always made that it would, if it were not for air resistance, but courses seldom deal with how a trajectory would look if air resistance *did* play a part.

The apparatus and its use is described in the following extract from my book Basic Engineering Mechanics Explained, Vol 2, pp 39 - 40.

The apparatus consisted of a spring-gun, triggered by switching off the current to an electromagnet that held the plunger in place against the compressed spring. The plunger struck a golf ball out of a machined cup, at a measurable angle α, with controlled velocity that gave it a maximum range of approximately 3 m.

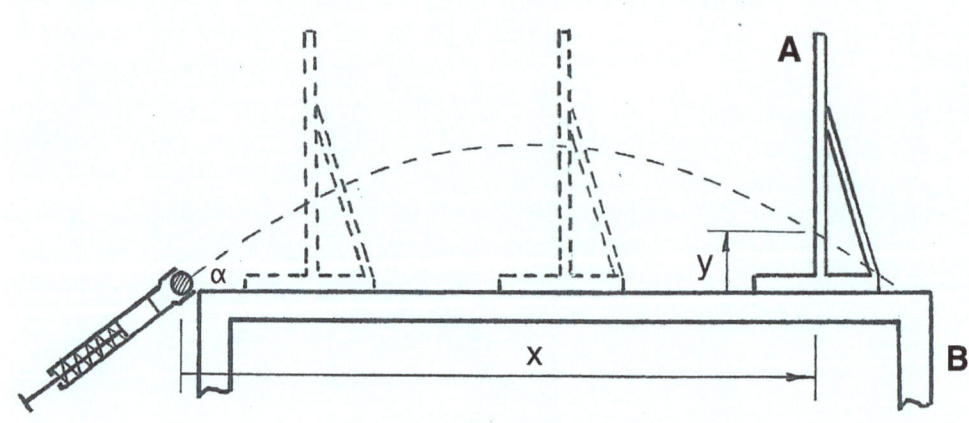

The spring gun was mounted at one end of a table, B, with a horizontal surface. Vertical panel A could be placed in various positions with known x-co-ordinates along the table.

The angle α was set and the ball was fired with a velocity that would not take it beyond the end of table B. The vertical panel A was covered with paper, over which a sheet of carbon paper was taped. Each strike made a mark on the paper, so that the vertical position of the ball could be co-ordinated with its horizontal distance from the launching point. Panel A was moved to different positions for successive shots, enabling the trajectory to be plotted.

Over many experiments carried out on many different occasions, by different groups, it was found that the trajectory for a launching angle of 45° fell short of a symmetrical parabolic trajectory by approximately 8% in the horizontal range. The symmetrical parabola was constructed using the co-ordinates of three plot points from the first half of the actual trajectory.

It is not hard to imagine the extent of the eager involvement of the students when using this apparatus. The same would apply to any experiment in which the motive is: 'Let's see what happens'.

Strategy 7: Conduct demo-labs.

If shortage of equiment or assistance is an issue, demo-labs can be very effective. This author conducted many demo-labs with classes of up to 60 students. It requires only one piece of equipment, which is set up in full view of all class members, and a few sets of measuring equipment like rulers and stopwatches.

The measuring equipment is distributed among the students, and when, for example, stopwatch readings need to be taken, each of the six students with a stopwatch reports his/her reading to another student who is deputised to write these readings up in a table on the board, to be seen and recorded by all present. Having to actually write them down focuses students much better than having the data distributed to their computers by electronic means.

Other students are assigned to process the raw data from these tables, and check one another's calculations. At all stages of the process, the instructor can give explanations, ask questions, and direct activities. All class members will come away with the same data to process, and all have the opportunity to ask questions that occur to them during the experiment.

If they subsequently have to write a report, they have no excuses that there may have been gaps in their method or findings, because the instructor would have made sure that everything was done correctly.

If one bears in mind that interest is generated by the requirement to use one's brain, one can direct proceedings in a demo-lab, as much as is possible, to:

- Minimise routine actions, and
- Pose challenging questions to the class.

Ways of getting students to reflect on lab experimentation

Post-lab tutorials

A tutorial could be arranged for the day following the conduction of a particular experiment. At this tutorial, one could issue students with a set of open-ended questions to answer through discussion in their groups. They would be told that some of these questions will appear in the lab test at the end of the term, so they would do well to make sure that they understand the answers.

Some questions that apply to all experiments, could include:

- What features of the apparatus could be improved, and how would you suggest this be done?
- Which readings were most difficult to take, and how would one overcome the difficulties?
- What were the nuisance factors you encountered, could they be eliminated, and, if so, how?
- Does this experiment 'prove' anything, and if so, what?
- Do you have confidence in the results you obtained?
- Can you see any way in which your conclusions could be challenged?

Examples of questions that apply to specific experiments and the apparatus used in them:

For instance, with an experiment to investigate the relation between the tight side tension and the slack side tension in a rope passing over a fixed cylinder:

- Why was it necessary to keep the rope moving during measurements?
- Why would you get a range of different readings when you repeat readings of S for the same value of T?
- How accurate was your measurement of the value of both these forces?
- Could you suggest a way of making these readings more accurate?
- Why was it necessary to plot a graph instead of simply recording the values of the ratio between a number of pairs of measurements of T and S?

Some questions could call for suggestions about original experiments or apparatus:

Even though students at a first year level don't usually have much experience of experimentation, they do have imaginations, and by allowing them scope to expose their imaginative thinking to peer feedback, they can learn a lot. The post-lab tutorial could contain questions like these:

- Was the string used for this pendulum ideal? Suggest characteristics that would be features of the ideal string, and explain your reasoning.

- The rolling resistance of the scale-model wagon was measured by determining the force needed to pull it slowly on a level surface. How could we be sure that this value remains constant at all speeds? Can you suggest a design for an apparatus that would enable rolling resistance to be measured for a range of wheel speeds, while eliminating the effect of air resistance?

- We determined the radius of gyration, **k**, of the rectangular slab, by two methods. Method **A** was to determine the mass moment of inertia of the slab by calculating its mass distribution relative to the axis about which it was caused to swing. Method **B** was to time its oscillations and use the value for the period thus obtained in the equation for the period of a compound pendulum. If the two values for **k** were not identical, does that disprove the validity of method **B**? How closely do you think these values should match to have confidence in the conclusion that the two methods produce the same result? Why?

Some questions that apply to the general process of experimentation, could be asked, for example:

What is the purpose of an error analysis?
Why does the conclusion of an experiment have to refer back to the stated aim?
Should all stopwatch readings be generally regarded as inaccurate, simply because there was a human operator?
When plotting a graph of data, where the independent variable can have any chosen value within the range of the apparatus, why is it unacceptable to join all the plotted points to make a zig-zag graph?
Is the 'result' of an experiment different from the 'conclusion'? Give an example.

Getting the most out of lab reports

Writing lab reports properly is an important skill. A lab report should contain all the elements of any technical report that engineers or technicians may be required to write in the course of their work. Learning to write credible lab reports sets the pattern for being able to compile research reports and theses.

Students often find it tedious to go through all the standard requirements of a technical report with the level of attention to detail that is required. They don't immediately understand the need for all the recognised standard parts of a lab report, so they gloss over the difficult ones, and often end up writing reports with excruciating shortcomings.

In order to impress upon students the reasons why experimental reports need to have the structure that sound engineering practice requires, an exercise like the following could be done *before* they are expected to write their first lab report:

In a tutorial, all the members of a class are issued with a photocopy of a lab report that exhibits several shortcomings, written by a student from a previous cohort. Without any previous guidance, students are given a certain amount of time to mark the sample report, and then to discuss in groups the criteria they used to assess the report.

Each group representative then has to explain to the assembly why their group allocated the mark that they gave: what the shortcomings were, and what they would consider to be more acceptable information than that presented in the sample report. A session like this brings home to them the reasons why reports on experiments need to be done in a specific way.

Having participated in such an event, students would be more likely to appreciate the list of criteria for acceptable reports that is subsequently issued to them.

When to require that students write lab reports

With my classes, over the years, I endured having to mark so many inferior lab reports that I eventually changed my system so that I would need to mark only one report from each student. That would be the one they would have to write in the lab test.

In the lab instruction booklet I suggested that each student should write up, but not submit, a rough lab report on each completed experiment, according to the guidelines given there about how a report should be structured. These reports would serve as study material for the end-of-term lab test.

There were two reasons why I stopped marking reports on every experiment that my students performed.

Firstly, it was frustrating to encounter the same thoughtless mistakes in report after report. Many students did not pay attention to corrections that I had indicated on their reports, and made the same mistakes repeatedly in subsequent reports. These students did not take the trouble to improve on their reports after their first one had been given feedback, because they were trying to put in minimum effort and still pass. They knew that as long as they kept on getting a pass mark, they were safe from failure.

If a pass mark for a report was 10 out of 20, some students were satisfied with getting 11 or 12 every time. They could end up getting a pass-mark while not understanding many of the most basic requirements that make a report scientifically credible.

Secondly, I found that the marks awarded to lab reports by my senior-student tutors were extremely unreliable, so I ended up taking over the marking, and found myself with way too many reports.

It was much more efficient use of my time to mark only those reports that students had to write in the lab test at the end of the teaching term. Also, because I was marking it, and not trusting the marking to one of my tutors, I was more confident that the marks were being fairly allocated, and consistent.

Lab Tests

These are written tests aimed at detecting the extent to which students have understood the principles of experimentation from doing the prescribed laboratory work. Various types of question can be asked, with the purpose of establishing whether the students:

- Have internalised the purpose and method of the experiments they were supposed to have done in the lab,
- Have understood the method of use and the possible shortcomings of the apparatus and the measuring instruments which they used in the lab,
- Know how to write a lab report, and why it is done in the standard way,
- Can process data that is given to them concerning hypothetical experiments that they are presented with.
- Are able to suggest modifications to experiments that will enable related phenomena to be investigated: variables that might not have been considered when they did their prescribed experiments.
- Can design *new* experiments to investigate relationships between variables that they might have been exposed to in their course, but that did not feature in their prescribed laboratory work.

Lab test questions would be really hard to get right if one had not done the requisite work in the lab prior to the test. Provided the questions are well-designed, a high lab test score indicates that a student has a sound practical grasp of the type of thinking that enables engineers to process the data that comes from experimentation.
Students could be given access to past lab tests to get an idea of what kind of questions to expect.

Some examples of questions that can be used in lab tests are given in Appendix 3.

Chapter 7

Using written assignments to promote learning

The way that the act of writing contributes to learning

The act of writing down a statement of what you think is true provides you with feedback. This process occurs in several steps, in which you:

- arrange your initial thoughts on the matter,
- find a way to express them verbally,
- physically perform the writing, whether by hand or by keyboard,
- read what you have written and judge whether that makes sense and expresses what you really think, and
- can modify what you have written until it makes sense to you.

Going through these steps leads to the consolidation of your understanding of the subject you were writing about.

There are two ways in which one's learning can be developed by having to write down statements about a subject, namely:

- The writer is compelled to engage with what is *generally* accepted as true, and
- Can develop his or her *own* understanding of what is true.

In the long run, your competence will be judged by the depth and breadth of your own knowledge structure. No engineer will be sought after by employers for merely being able to quote a series of generally accepted 'facts'.

Learning activities really ought to prompt students to examine their state of understanding, not just to replicate what they have been told. This can be accomplished by requiring students to answer questions in writing, in their own words.

In some fields of study, this aim is customarily exercised by requiring students to write essays. In other fields of study, including engineering, students are sometimes required to provide short definitions and explanations as part of their assessments.

However, tasks like these are seldom deliberately used as developmental exercises outside of the context of assessment. I believe they can and should be used as learning activities in the study of engineering mechanics. This is how:

Short-answer questions are ideally suited to being used as stimulus material in a tutorial.

A 'short-answer question' requires a written answer that could be as short as a single sentence, or run to two or three paragraphs. It might require a sketch and a brief mathematical explanation. It asks the students to explain some aspect of the science of mechanics in their own words.

Why in their own words? Because 'own word' statements are the most authentic means of deducing whether or not someone understands what they are talking about. Having to undertake such a task compels students to organise their ideas about the topic. Having to do it in a small group that has to come up with a communal 'best' explanation leads to students seeking clarification from their peers.

Some examples of short-answer questions to use in a tutorial:

- Define 'mechanical work' in a sentence or two.
- Explain why the amount of work done to compress a coil spring by a distance 'x' is proportional to x^2.
- What is meant by doing work *against* a force?
- Sketch a graph of force vs extension for the force exerted by a bowstring on an arrow, from the position where the arrow is just released from full draw, up to the position where the string is again taut.
- Explain the principle of operation of a ballistic pendulum.
- Explain what is meant by an 'impulse' and an 'impulsive force' in relation to momentum.
- Give reasons why it is not possible to predict how much energy will be lost in a collision, without empirical testing.
- Describe what is mean by a 'coefficient of restitution' and state its uses.
- How can one measure the mass of an object without weighing it?
- Under what circumstances is the velocity of an object identical with its speed?
- If the Newtonian view of the physical universe has been made obsolete by the

scientific theories of Einstein and others in the 20th century, why is it still used by engineers?

It is not much use issuing short-answer questions as a take-home assignment for the purpose of preparation for a tutorial, because the answers could be plagiarised, spat out by AI, or taken from the definitions supplied by more knowledgeble people.

Short-answer questions that are included as part of the previously unseen stimulus material for a tutorial definitely provide a context and purpose for discussion.

If a sub-group has to come to a unanimous agreement on the best way of answering a short-answer question, the discussion that occurs will ensure that some consolidation of understanding takes place.

How useful are short-answer questions in assessments?

Some instructors avoid including them in assessments, on account of the concentration required to mark them. Poor handwriting, deficient grammar, and incohesive logic make it really difficult to stay patient while marking some students' work.

I always included a small set of these questions in a formal assessment, because the answers that one gets are very revealing. Particularly, the answers to questions requiring a simple freehand sketch or diagram.

My reasoning for setting such questions is that in order to be a competent engineer or technician, you have to be able to communicate technical ideas to other technical people, verbally and in writing, and with the occasional use of a freehand sketch. In the workplace, your competence is never going to be assessed purely on your ability to perform mathematical analyses. People are going to assess your competence according to how you express your ideas.

My book 'Revision Exercises in Basic Engineering Mechanics' provides 242 short-answer questions, covering almost every aspect of basic mechanics. These may be used without requesting permission.

Longer written assignments for individual students

It is standard practice in various academic departments to issue assignments requiring some research. Essay assignments, for example, are beloved in the social sciences and the liberal arts. Is there any sort of useful equivalent for students taking a course in mechanics?

To my knowledge, colleagues teaching engineering subjects have at times issued 'essay' equivalents for submission. These include speculative design projects, lab reports, descriptive reports on topics that the students had to research from reference sources, and reports that detail the procedure followed in a design-and-build project.

Whether or not to include this type of exercise in a first course in mechanics depends on two factors:

- What you think that doing the assignment will contribute to the students's grasp of the subject, and
- The logistics of marking.

Some of the aims of issuing longer assignments for submission include hoping that students will:

- Broaden their appreciation of the applicability of what they have been exposed to on the course, and
- Exercise the skills of research and reporting.

These are lofty ideals, whose value cannot be denied. However, who can say if the activity actually results in the outcomes imagined by instructors?

Reasons why long written assignments might not be satisfactory forms of engagement with the subject or result in fair assessments include:

- Students may not have enough time to do justice to the task,
- Some students work to a pragmatic short-cut policy of putting in minimum effort, so they weigh up how much the assignment is likely to count, before deciding whether it is worthwhile to devote any time to it at all.
- Students might be just plainly not interested in the topic you assign.
- Some students appear to think that presentation skills are simply window-dressing and not something that engineers need to be bothered with.
- Students can present the researched material without internalising it. They look it up, write it down dutifully in fulfilment of what is asked, and then forget it.

I personally issued longer written assignments only when my classes consisted of fewer than thirty students. In other circumstances, when I had three classes of approximately 90 students each, I wouldn't even consider using them.

However, supposing the logistics permit, and one has sufficient marking support to issue research assignments:

In that case, if you do issue such an assignment, it is a very good idea to ensure that each student chooses a unique topic. This is so for two reasons: firstly, it minimises the likelihood of copying and work-distribution. Secondly, and more importantly, students will put energy into whatever they find interesting. If they can choose their own topic, they are more likely to tackle it with enthusiasm.

For assessing longer assignments carried out away from a controlled classroom environment, one can never be sure whether the work handed in really was the unaided work of the student.

In this author's view, when a student tackles an assignment, what matters is how much they learn from doing it, not how complete and convincing it looks.

For the purpose of getting learning to happen, it shouldn't matter how much help the student obtained, or from whom, in the course of doing the assignment. By obtaining help from anyone or any resource, a student often learns something useful that would not have been covered in the course of study. In fact, they would even be likely to learn a few things that their instructors don't know.

Once I realised this, I encouraged my students to ask for advice from anyone they could find who could add to their knowledge about the topic, whether senior students, family members, technicians or professional engineers. Any broadening of experience of this type I regard as essential in the education of an engineering student.

However, the only way to assess what the student *has actually learnt as a result of doing an assignment*, is to test that knowledge under examination conditions *after* the submission of the assignments.

Naturally, If every student has done a unique project, the logistics of doing this would be difficult to arrange.

One way around this is to set a surprise test within a few days of the assignments being submitted. In that test, students could be asked to write down a summary of what their personal assignment had investigated and had concluded. This summary would need to be be kept to a specified number of words, say 300, and would be marked according to its credibility, conciseness and success as a communication.

Group assignments

In engineering departments, group assignments are commonly used for information-gathering, experimental research, design projects, and design-and-build projects. Sheer force of student numbers drives lecturers to issue assignments that need to be tackled in groups. The understandable reasoning for this is: if you have 250 students and they work in groups of five, you have only 50 assignments to mark. That's pragmatic, and it's fine.

Putting five students together is also supposed to fulfil another often-stated aim of group assignments, namely to compel students to cooperate, as cooperation is something widely held to be desirable. The reason that cooperation is advantageous for the engineering undergraduate is that it provides the opportunity to associate with, and get feedback from one's peers while one is still getting to know what is expected of a student.

The need for cooperation is not a holy grail in itself. It doesn't suit eveybody. If you turn out to be a great engineer or scientist like Archimedes, Nikola Tesla, Galileo or Isaac Newton, when no-one around you is in your league, you would just be held back by being forced to cooperate.

As in family life or business life, some people cooperate nicely with certain other people and reluctantly, or not at all, with certain others. So, it is definitely not a given that five students placed in a group will work together willingly or efficiently.

It happens frequently that some students split from the group to work on their own, frustrated with the disorganisation evident when the group meets. Students quite naturally resent the lack of useful input from lazy group members and don't want to be dragged down by other people's poor performance. Personality clashes and leadership issues bedevil attempts to work in a group. It is no use saying 'You must learn to work together, because the ability to co-operate is a desirable skill'. Some people just won't. Which brings us to the next issue:

Awarding fair marks for group assignments

While I approve of group assignments, I do not believe it is reasonable or fair to award every member of a given group the same mark.

Frequently, some group members have contributed more than others to the assignment. I sympathise with the high-contributors. It is simply not fair that slackers get disproportionately high marks for their paltry contributions.

The only way around this problem is to assess each student's work individually.

One could allow the project to go all the way to completion as a group effort, but shouldn't consider the mark for the group effort as final.

Rather, set a separate test, after the submission date, based on the project, that all participants have to take *individually*. Those who contributed and understood what the project was about will do well, while those who cruised will not. If desired, the mark for this test and the mark for the group effort can then be combined.

A marking method to avoid

Some of my colleagues tried another method, with generally good intentions, but the results were problematic, so the method was dropped.

These colleagues decided that they would allocate a preliminary mark to a group assignment that had been submitted, but would not publicise this mark until they had taken it to the following stage:

They asked each student to complete, in confidence, a form on which they ascribed a number out of ten to the effort put in by each of the members of the group. So, if you, as a student member of a group, think Jimmy did hardly anything, while Charles and Sandra put in a huge effort, and you yourself did a reasonable amount, you might answer Jimmy 1, Charles 9, Sandra 10, and me 6.

My colleagues thought that processing the group mark for the assignment with a factor based on these numerical allocations would lead to a way of giving each student a fair grade. That method was dropped pretty smartly when they realised that almost all students were inclined to overrate their own contribution, whether innocuously or with devious intent.

Students are not alone in this: university instructors fall prey to the identical vanity. I once read an academic paper describing a survey in which all of the lecturers and professors of a particular department were asked to assess their own ability as a teacher. It turned out that 96% of the respondents thought they were in the top 25% of the teachers in their department!

Naturally, whichever method you propose to use for assessing an assignment must be spelled out in advance, at the time of issuing the project brief.

The kinds of topics that are suitable for written assignments

On a course about the principles of mechanics, it would not be productive to issue topics that are merely of general interest to engineering. Assignment topics should deal with the principles being studied on the course.

There are two general classes of assignment: active investigations and researches based on consulting reference material:

1. Active investigations

From a learning point of view, the most effective written assignment is one in which students report on something that they actively investigated by physical experimentation.

The focus could be an experiment that the students conducted on a mechanical topic outside the scope of the offical lab programme: For example:

- An attempt to replicate the Cavendish experiment,
- A comparison of the merits of two mechanisms that were supposed to accomplish the same results, for example the performance of two methods of propelling a human-powered water-craft, or
- An analysis of an experiment that has been proposed but not yet performed:
- Reverse engineering of, and improvements tested on a machine or mechanism.

Students should be reminded at the start of this kind of assignment that the written report is to be focused on the actions taken, the findings and a discussion of the mechanical principles involved. While it is relevant to show very briefly the historical context of the topic, this type of assignment should *not* consist mainly of reference material.

2. Presenting material researched from references

Provided there is time to include them, assignments like these would be meant to broaden a student's appreciation of the place of the science of mechanics within the broad scope of engineering.

Typical examples:

'The history and development of mechanical speed governors for engines';
'Ergonomic constraints on the design of fairgound rides';
'The history and design of treadles for operating rotating machinery';
'Historical developments in mechanical methods of pumping water';
'Methods of raising heavy loads in Roman times';
'Techniques of gear construction in the middle ages';
'One significant machine developed during the industrial revolution', and
'The efficiencies of different types of waterwheel'

Such assignment topics would be motivational for those students who are sufficiently interested in engineering. However, students would need to be convinced that taking on this type of assignment would be worth their while. Could they be persuaded of the gains to be had, both in learning and in credits?

The chief purpose of such an assignment would be to give students practice in consulting resources and summarising the information that they so glean. While both are important skills in the development of an engineering student, assignments of this type might best be left for more senior courses that examine the place of the engineer in society.

Chapter 8

Predicting the outcome of an operation of a real mechanism

In this type of learning activity, students have to predict a mechanical outcome that can be tested in reality. Participants receive a score based on how close their prediction comes to the actual result. A suggestion for the method of scoring all such exercises is outlined at the end of this chapter.

For each exercise of this type, a piece of apparatus has to be specially constructed. Generally, it will need to have one or more moving parts that can be made to interact to produce a result. The apparatus should be prevented from being operated until the instructor is ready to conduct the reality-test. Prior to that reveal, students are allowed to inspect the apparatus and make whatever measurements they think are needed, in order to predict the outcome. For example:

Exercise 8.1

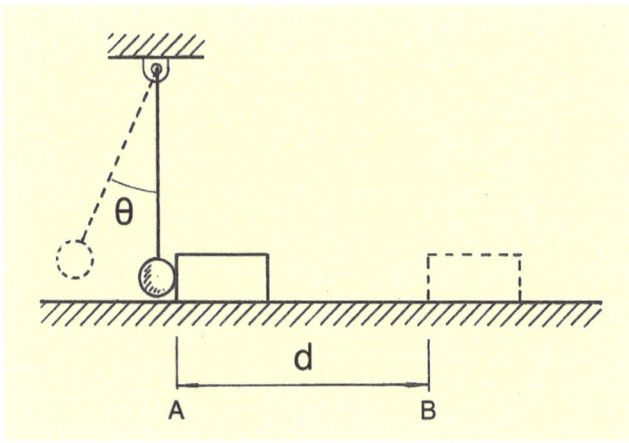

The pendulum with its mass-piece could be locked into a position from which students cannot move it, so they can't operate the apparatus before the instructor does the demonstration.

With this apparatus, the task is to predict distance d,

namely how far the block will move before coming to rest, after being struck by the descending pendulum, given a nominated value for angle θ, the angle from which the pendulum is allowed to fall.

Justification for making use of such exercises

There is no point in being slavishly attached to any theory that doesn't match the reality it purports to describe.

Taking part in an exercise like this challenges students in a way that no theoretical exercise could. In a test or exam, any student's answers could be right or wrong, and no-one but the marker would know, or care. However, in this type of exercise, their knowledge of mechanics is on show for all their classmates, and whoever else is present, to see. Metaphorically speaking, each of them would be called upon to put their money where their mouth is.

A variety of suitable pieces of apparatus can be designed and built. Below are presented several suggestions for different configurations of apparatus, and the way in which they could be used. Hopefully there will be a place in which to store them, because they are likely to be used on many occasions.

Exercise 8.2

A straight scale-model roadway with a dip in the vertical plane is constructed, with dimensions similar to those shown below.

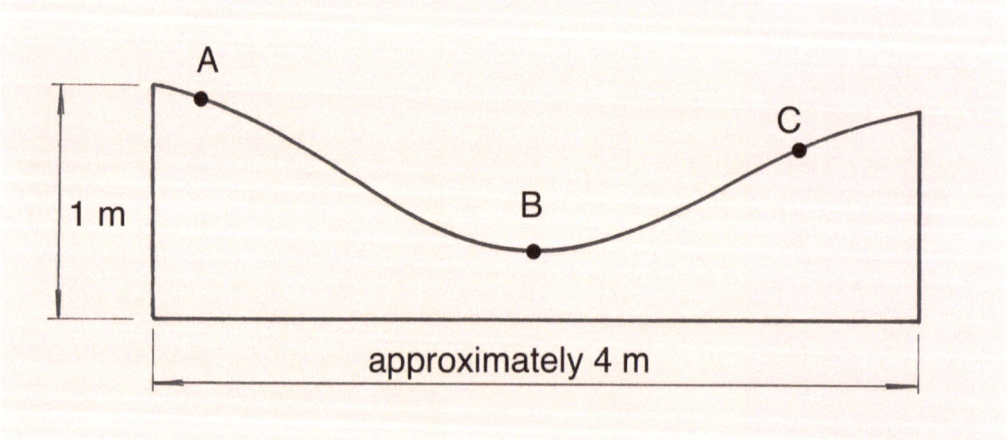

A small purpose-built wagon with four wheels is allowed to roll from rest at a nominated point **A**, through point **B**. It will come to rest momentarily at some point **C** before rolling back. Immediately after the wagon starts rolling back from point **C**, the instructor puts a stop in place to prevent the wagon moving further towards point **B**.

The task, from that point in the proceedings, is to predict how far up slope **BA** the wagon will roll on its return before coming to a momentary stop again.

Once all the students' individual predictions have been recorded by the organisers, the wagon is again allowed to roll from rest at point **A**, and this time it completes the motion. The position of the momentary stop on slope **BA** is recorded, and the predictions are graded according to their accuracy.

For successive uses of this apparatus on subsequent occasions, the position of point **A** could be varied, and a different wagon, or one with different wheels could be used.

Exercise 8.3

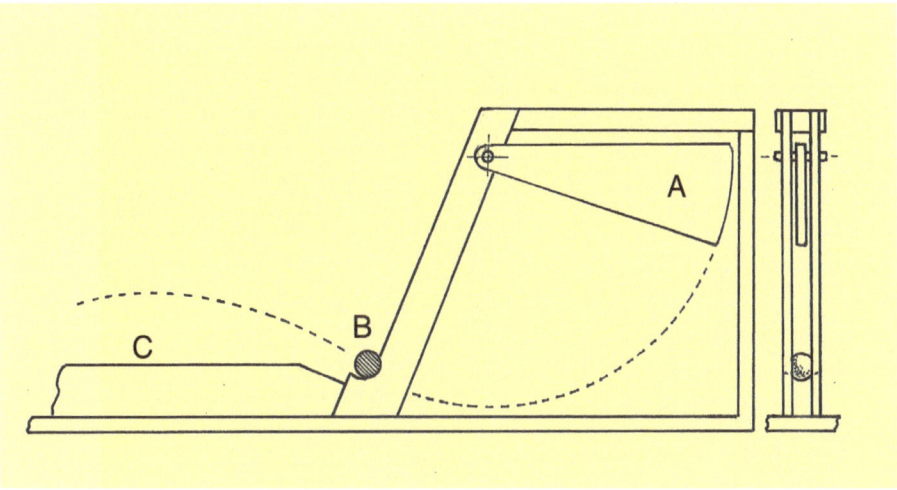

Pendulum **A** gets released from rest, and swings down to knock ball **B** out of a 'cup', projecting it into the air over a long base plate **C**. Participants have to predict how far the ball will fly before hitting the base plate. The position of the strike point on the base plate is established by having the base plate covered in paper, with carbon paper on top of that.

For variations, pendulum **A**, shown here as a sector of solid material, could have extra mass added to it, or could be replaced with a solid rod or a rod with a ball at its end. Ball **B** could be replaced with balls of different material or size. For further variation, a different length of pendulum could be used, pivoting about a different point.

Exercise 8.4

This one requires students to predict the time needed for a rotating axle to come to rest.

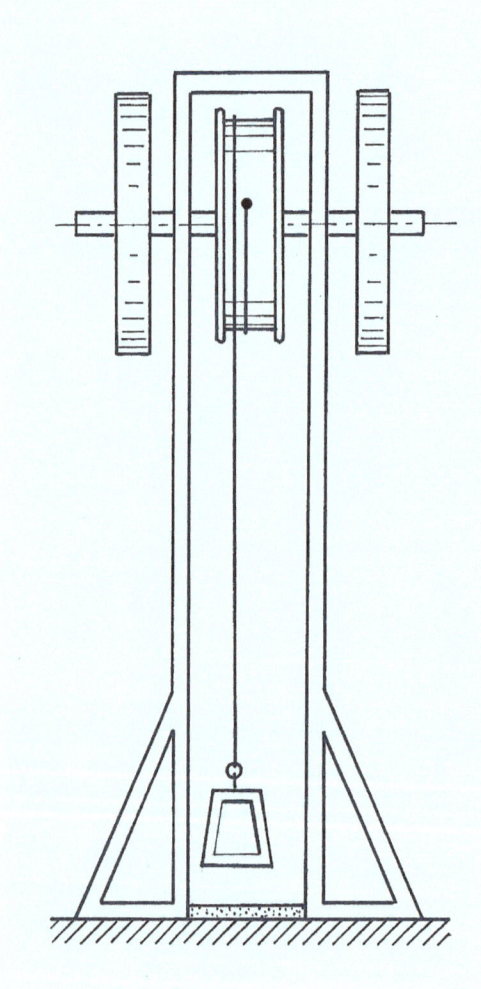

A horizontal shaft is mounted in bearings in a frame. To this shaft is fixed a central drum. Additional discs may be fitted to both ends of the shaft, to vary the mass moment of inertia of the rotating assembly.

The drum is constructed with a small radial hole in its circumference, so that the end of a cord can be pushed into the hole. The rest of the cord is wound around the perimeter, with a mass-piece attached to its further end. The length of the cord is chosen so that the loose end of it comes out of the hole just as the mass-piece hits the rubber base plate.

Stopwatches are started when the mass-piece hits the base plate, and stopped when the shaft has come to a stop.

To provide consistency about the exact moment when the axle stops rotating, the instructor can nominate someone (not a competitor) to strike a bell at the instant at which that person judges the axle to have come to rest.

Exercise 8.5

A solid metal cylinder is welded to a compression coil spring, The assembly fits loosely inside a clear perspex tube. The assembly is dropped from height **a** and bounces back to height **b** on its first bounce. Contestants must estimate the value of height **b**, given the dimension **a**.

The reason for taking the height measurement from the top of the solid cylinder rather than from the bottom of the spring, is that the spring will most likely still be oscillating when the assembly reaches its maximum height.

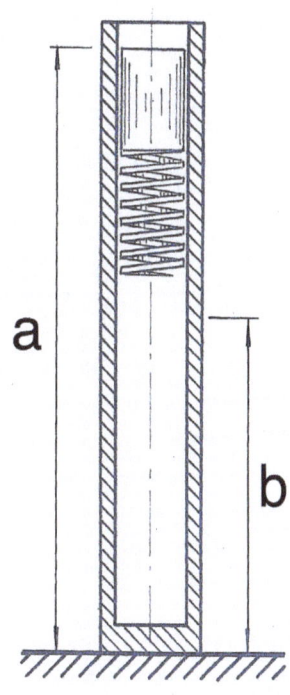

Exercise 8.6

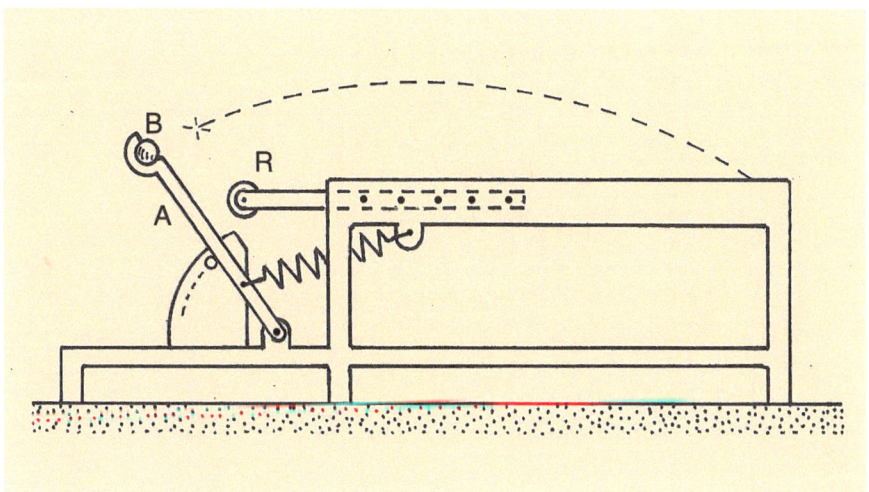

This one is a variation on Exercise 8.3. In this case, arm **A** is pulled back against the spring, and released to project the ball. The requirement is to predict the distance to the strike point of ball **B** on the table.

A shock-absorbing rubber roller **R** defines the release point of the ball, which could, for convenience, be a golf ball.

Some variable quantities are: the position in which roller **R** is set, the position of the stop on the quadrant, and the spring constant of the fitted spring.

Exercise 8.7

A beach ball is pumped to a measurable pressure. Participants are allowed to measure and weigh it, without removing it from the room in which it is presented.

The task is to predict how much mass should be added to a submerged weight hanger suspended from a fine nylon net over the ball, in order to submerge the ball just below the surface of the water.

An additional factor to be considered (which should not be pointed out to participants) is the effect of buoyancy on the mass-pieces and the weight hanger themselves.

The test could be conducted with the apparatus in a water tank or swimming pool.

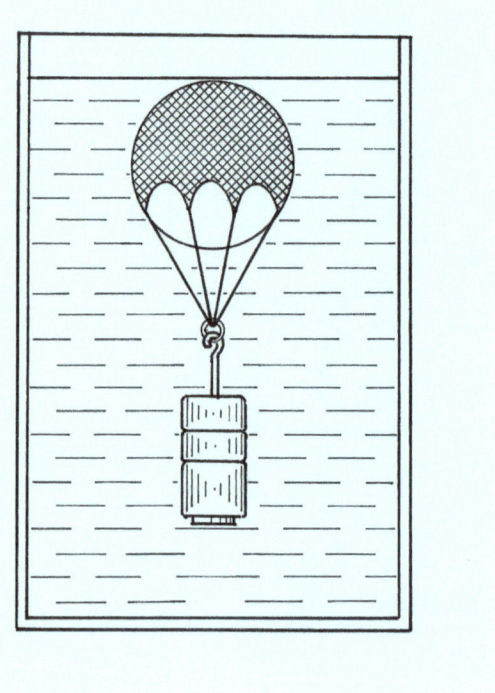

Suggested scoring method for prediction exercises

Once the actual result of the operation of the mechanism has been established, each person's prediction of that result needs to be allocated a score. Your score must be a measure of how closely your prediction matched the actual result.

See diagram that follows:

Illustration: Suppose candidates had to predict the distance that a golf ball would be projected, using an apparatus similar to that in Exercise 8.6 above.

If the prediction that was furthest from actual result was by student D, with a deviation of 804 mm, and the prediction by student A had a deviation of 557 mm, then

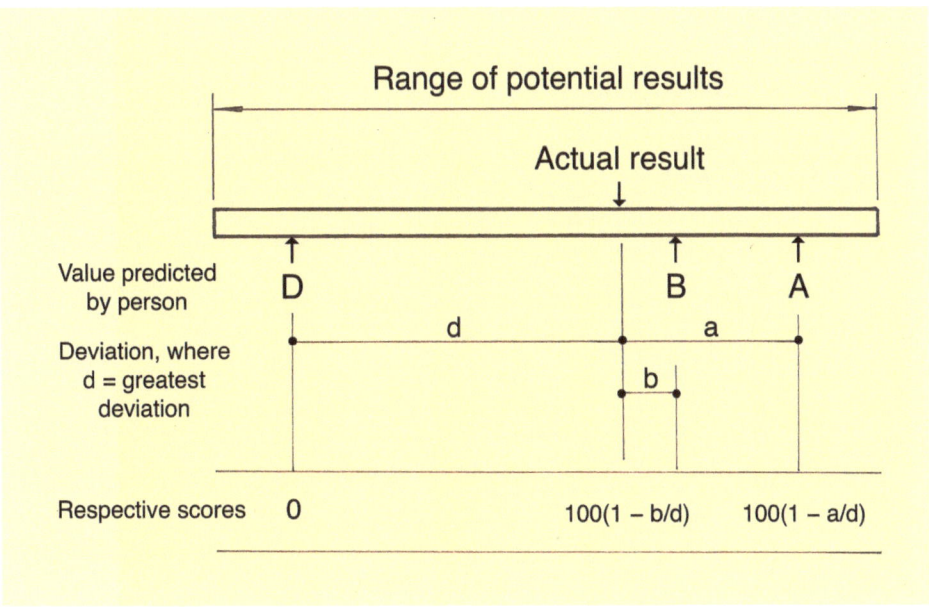

Student D would score: $100[1 - 804/804] = 0\%$

Student A would score: $100[1 - 557/804] = 31\%$

If the deviation of student B's prediction was 125 mm, then B would score: $100[1 - 125/804] = 84\%$.

If another student C predicted the value that matched the one subsequently obtained in reality, namely having a deviation of zero, then C would score

$100[1 - 0/804] = 100\%$

Predictive exercises like these can be used to create a competitive occasion called a 'Predictathon'

This would be an event during which candidates have to make predictions for the operation of several different mechanisms of the type indicated in the above examples.

Any given student's performance would be their aggregate score for their predictions on all the pieces of equipment present.

A predictathon could be used as one of the set of assessment events on the course.

It could also be used as a socially competitive event that will increase the public visibility of the department. For instance, as a competition amongst a group of colleges, or students of one college versus engineers and technicians from local industry. Students in high school could also be invited to participate.

Taking part would be motivational for any interested persons, irrespective of the extent of their training. This type of occasion is bound to motivate participants in several ways, namely:

- By giving them the chance to identify with the department in a social context, similar to what happens when taking part in a design-and-build contest on testing day,
- By pitting them against a problem that is based on reality, rather than one based on abstractions,
- By challenging them to pull together everything that they know about mechanics, expecting them to be at the top of their game,
- By giving them public recognition for their skill in taking many factors into account in order to solve a mechanics problem.

Chapter 9

Adapting lectures to ensure that students are active, not passive

How effective are conventional 'teaching' lectures?

Lectures can be used for different purposes: to inspire, impress, impart information, entertain, and to challenge. Those who are really good at lecturing might accomplish all of these purposes in any given lecture.

To some extent, certain lecturers can get away with this type of one-way flow of information, on account of their personal charisma. If you have an entertaining personality, of course your lectures will be popular. Such lecturers tend to get higher ratings from students than do others.

However, one should not confuse the entertainment value of a lecture with its successfulness. The important issue here is: how effectively do these lectures result in learning?

The 'teaching' lecture was once sardonically defined as 'An occasion on which the lecturer's notes are transmitted to the students' notes without passing through the minds of either'. That kind of lecture predominated at the university where I was a student. Some people still lecture in this way, unaware of the possibilties of making better use of their students' time.

Lecturing appeals to some instructors because it is an efficient way for *them* to use *their* time. They know their stuff, and they can easily construct a sequence of logic to present what they wish to convey. If you have to teach something to a large group, especially at short notice, one very effective way that you can use *your* time to prepare for it, is to arrange your own already-well-organised thoughts about the material into a sequence that can be lectured. However:

Is a lecture an efficient way for *audience members* to make use of *their* time?

Consider what you, as a student, would be expected to do during a monologue teaching lecture. Namely: to concentrate intensely for 50 minutes, during which time you have to

- Accept your passive role as the non-participating receiver,
- Take notes at speed,
- Make an effort to understand what is being explained,
- Attempt to fit any new revelations into your existing view of the subject,
- Split your attention between what the lecturer is saying and what he or she is writing or showing visual images of, and
- Suppress questions that arise in your mind whenever you missed a point, thinking you will ask these later, if there is time, but there seldom is.

At the end of such a lecture, as a recipient of it, all you would have gained is a set of incomplete impressions of what was 'covered'. Then you have to fill in the gaps by additional reading, consulting classmates, attempting exercises issued to you, or arranging a meeting with the instructor.

Even if there were no gaps in your recording of what was presented, all you would be left with is a summary of the knowledge-structure paraded by the lecturer. What can be done with this? The obvious answer is to swot it up and replicate it in the exams. If replication is the aim, it doesn't say much about the college's claim to providing an education.

If the lecture content had been presented in print form, you could probably have read it over twice in fifteen minutes, and would have had another half hour available to make some of your own notes about what was important to remember.

However, even that option would be unsatisfactory. What would all this focused individual study have done for your motivation levels, your interest in the subject, your confidence in your own grasp of the topic?

Perhaps the most dismal aspect of having attended such a lecture is that you were expected to absorb this material whether or not it meant anything to you. For the sake of passing the course, you might have needed to know this stuff, but: did you *want* to know it?

How important is 'wanting to know' about a topic?

Consider this experience related to me personally by Jack de Wet, a former Dean of the Faculty of Science at the University of Cape Town, who, as a young academic, had attended a seminar at Princeton, given by Einstein:

Jack described how he was one of a group of young physicists who were seated in awed silence, well before the time the seminar was due to start, eagerly awaiting the presence of the great man. Eventually, in walked Einstein. He looked around the room and said 'Well, gentlemen, are there any questions?' The audience was stupefied. No-one said a word. Einstein responded: 'In that case, I will see you tomorrow,' and out he went. The next day, all the participants had done their homework. Questions abounded. The seminar had sprung to life.

Einstein clearly knew that you can only learn something if you are ready and eager to learn it. So, ideally, what you explain should be based on what the students are curious to know. Anatole France once wrote that *in order that knowledge be properly digested, it needs to have been swallowed with a good appetite.*

Issues around conventional lecturing

In the 1980s there was a well-meant short craze in the academic teaching world that some people thought would spread enthusiasm for their subjects to many other classrooms. They made videotapes of photogenic, entertaining and knowledgeable lecturers giving conventional lectures. These tapes were shown to students in successive cohorts and in other colleges. This was done in the hope that somehow, all that high-powered expertise and charisma would result in greater learning.

Not surprisingly, 'talking head' videotapes were a failure. It soon became clear that a 45 minute monologue is difficult to endure, cannot be interrupted by questions, and doesn't fit in with the unique relationship that develops between the local instructor, his or her students, and what they all want out of the experience. As a teaching method, this one was very short-lived.

If anything, the 'talking head' experiment showed us, by default, that interaction between students, instructor and subject is a key factor in the effectiveness of a lecture.

There definitely are features of 'conventional' lecturing that can be found valuable by students.

One of these is the appropriate use of anecdotes. As an audience member, you give your full attention instantly when you hear words like: 'One day, when I was working at ABC labs, a mistake like that caused an explosion in the factory...' This works because, by listening to the story, those in the audience now have the chance to use their imaginations to reinforce their own knowledge structure, as opposed to passively absorbing the knowledge structure very kindly offered to them by a well-meaning lecturer.

Another advantage that can be found in conventional lecturing is the inspiration that derives from the lecturer's own enthusiasm for the subject. In my first year as an engineering student, we had a boring chemistry lecturer. I couldn't get my head around chemistry, and developed an unhealthy dislike for the subject. A month or so into the term, our lecturer was replaced. The new man was young, a strict disciplinarian (with a class of 300 first-year engineers, you had to be) and full of fire. He knew his stuff, he put it across dynamically, and he was very obviously fascinated by chemistry. In a few weeks he turned my attitude to the subject from distaste to grudging interest. I realised then that chemistry was not boring. In fact, there are no subjects that are inherently boring, because for every subject one could think of, there is someone so interested in it, that he or she has made a career out of it. There are only *engagements with a subject* that are either boring or inspiring.

One can be inspired in a lot of different ways.

When I was a consultant in teaching methods at the University of Cape Town, they had an annual 'Distinguished Teacher' award, given to three academics each year. One year I asked the recipients of the award if I could sit in on one of their lectures, to see for myself why their students had nominated them. They all consented.

All three of these lectures were good, but for completely different reasons. Here is a comparison of two of them:

A zoology professor gave a textbook-perfect delivery, everything well-paced, clear, with good visuals, and designed to allow the students to make good notes and understand them. He welcomed questions at any stage, and answered them succinctly and with clarity. He really wanted the students to feel at home with the basics of zoology, and to find it easy to master. His motive was to start first-year students off with a positive attitude to his subject.

In stark contrast, a lecturer in the English department did everything wrong from a presentation point of view. Her 'lecture' was a monologue, not involving the participation of the students in any way. If she wrote on the board occasionally, it was jumbled, half written over previous things she had just written, in lines at odd angles, with spidery writing too small to be read from halfway up the room. However, she raved on about T.S. Eliot's work so eloquently, that I left there feeling: 'Wow, if Eliot's poems do this to her, then I definitely want to go and read some today!'

What these lecturers had in common was an intense love of their subject, and it showed. The main thing that you teach is not your subject, but your personal 'take' on the subject, and your attitude toward it.

Adapting a 'lecture' to foster curiosity and involvement

A competent instructor can find ways of heightening students' curiosity. It may seem daunting to have this as an aim, when you feel obliged to hold the floor with a class of 250 undergraduates that have to be taken through a fixed curriculum in a limited amount of time. Especially so, when the students are accustomed to being spoonfed with easily digestible information, and are expecting you to oblige.

However, if you have to take charge of a timetabled period called a 'lecture', in which you have a large group in one venue, you don't have to be the only speaker, and you are not obliged to present information non-stop for the duration. You can still use the occasion to make sure that students get involved.

This can be done by adapting whole or part of the event to become:

- A tutorial, with some stimulus material that you present, as described in Chapter 1,
- A demo-lab, as described in Chapter 6,
- An occasion on which students take turns to report to the assembly something which they have been researching, or
- A session in which you undertake to answer questions posed to you by students.

Two methods of getting students to generate questions:

Method 1: Tasking students with bringing to class any questions they wish to put to the instructor. Even if not all students are confident or diligent enough to come up with such a question, provided there are three or four worthwhile questions from the audience to kick off proceedings, the event usually turns into a high-energy debate, with great levels of participation.

I have personal experience of how well this works. One day, while I was a consultant on teaching methods, a colleague in the Education Faculty asked me if I would give a lecture on 'student evaluation of teaching' to his class of about 60 postgrad students in two days' time. I was happy to oblige, but very busy with other duties, and did not have time to prepare. I told him I would do it, on condition that we based the session on questions about the topic that his students brought to the class, so that I could talk spontaneously.
From the moment the first question was asked, the discussion was extremely lively. It was the best 'lecture' I had ever given up to that time. All I did was answer questions, with short explanations from my experience, when warranted. Every answer I gave prompted further questions. It made full use of all my experience in the field, and it was based throughout on what the audience wanted to know.

Method 2: Hold a quiz-based activity, in which sub-groups put to the instructor any questions they still have after finding out what they collectively did not understand in the quiz.

The following example of how this works also comes from my time as a consultant: on this occasion I was assisting a law lecturer who had asked for help in improving his lecturing method. I suggested he try a method I had found in the literature. So he tried this method, and I was there to watch. He told his class of about 130 3rd-year students that he was going to give only a short outline of the day's topic, for 20 minutes of the 50 minute period. He asked them not to take notes in the conventional way, but to jot down notes about anything that his presentation had seemed to gloss over, or whatever they felt needed clarifying.

After giving his twenty-minute outline of the topic, he asked them to form groups of four, two students sitting next to each other, with the two students in the row behind them. The members of each group were asked to pool their notes and come up with one question which they could put to the lecturer in front of the assembly. There was about 10 minutes of very intense discussion among all the group members, following which the lecturer gave the spokesperson of each group a turn to put their question to him.

He answered their questions until just before the bell went. When he saw that the time was nearly up, he asked those 12 groups whose questions had not yet been aired to write them down and bring them to him before they left the room. He promised to start the next day's lecture by answering these questions. Which he did. That next lecture was a sellout: full attendance, very unusual for first period. Those students already had a stake in the lecture before it began. The lecturer was delighted with the effectiveness of this activity.

So, while you might have a timetabled period called a 'lecture' for administrative convenience, you could adapt it to include a lot of opportunity for students to have their curiosity stimulated and to use their brains.

Conclusion

Selecting the right mix of exercises to use in teaching the subject of engineering mechanics

In this book I have assumed that the instructor has free choice in all aspects of how to run the course. I know that in some institutions, this is not the case. There are universities and colleges in which the department appoints a committee to prescribe required reading and to specify methods of teaching and assessment. Such an approach is well-meant in the sense of trying to ensure that standards are maintained. However, in this author's opinion, a policy like that stifles the initiative and creativity of instructors. After all, the most important ingredient of good teaching is the ability to motivate the student. Not much motivation is going to follow from classes that are run by instructors who are regarded as mere functionaries.

If you are lucky enough to have some say in the matter, choosing which learning activities to use in your teaching depends on several factors:

- How much time you have
- How much help you have
- How many students you have
- What venues you have
- What resources you have
- What you feel comfortable with doing
- What you believe your role should be
- What your students are capable of
- What outcomes you want for them, and
- What kind of assessment standards are demanded by your teaching institution.

As described in this book, there are many options for instructors, among the possible types of learning activity to use in teaching the subject of mechanics.

The theme that runs through these varied exercises is getting students to use their minds, rather than to be passive absorbers of someone's explanations.

It should be clear by now how this author came to the conclusion mentioned in the preface, namely that: *What you get your students to do* is more conducive to their learning than anything *you do* in front of them.

What about online teaching?

The reader might have noticed that nowhere in this book have I mentioned the use of so-called 'interactive' learning programmes, or 'learning management programmes'.

This omission is deliberate.

Activities and software like that have been designed to attempt to solve problems that were created by teaching institutions in the first place. Problems like: too many students, too many other duties expected of academics, not enough resources, the high expense of staff salaries and a misplaced wish to 'standardise' everything.

I have personally never needed to teach online. However, I have sat and observed an engineering instructor 'tutoring' a small group of students online.

These students were in separate locations, in regional towns, away from the city. They came to a common venue for a week, twice a semester. In between these occasions, they did not see a member of the teaching team, nor did they see any of the other students.

Working alone, they tackled exercises provided to them online. The only human feedback they got came from the weekly online 'tutorial' session, when they could talk to the tutor via their computers. By the same means they were also able to talk to the other students, but had little cause to do so. At this 'tutorial' there was no new stimulus material, and no set task that required students to converse with other students. It was simply an online version of a type-B tutorial, with all of the drawbacks mentioned in Chapter 1.

I sensed the frustration felt by the students on account of the inefficiency of the process, despite the fact that the instructor was well-informed, patient and helpful.

The students were not very forthcoming, possibly hesitant to say anything to the instructor that revealed gaps in their knowledge. Altogether, despite everyone's good intentions, this event did not make optimum use of anyone's time.

Learning occurs meaningfully only in a social context. Humans need to be inspired to like a subject. They benefit from learning in the company of others, in collaboration with others, and to have their efforts noticed and appreciated. You get a sense of your place in the world when you can measure your achievements against those of other people. It is simply not the same to sit alone, practising routines while engaging with a machine, or to be assessed by a machine. Even more futile, to be given pre-programmed feedback messages by a machine.

No matter how much the enthusiasts for using electronic technology in their teaching talk up their 'solutions', or how how likely it is that your students will need to use technology for the technical aspect of their work when they are practising engineering, mastering the subject of mechanics is about grasping the reality. No collection of algorithms, formulas or programs can be a substitute for experiencing the way the physical world works.

Then there is the human side of inspiration, which is absolutely crucial: the motivational ambience provided by the character of the teacher.

Broudy and Palmer, in their 1965 book 'Exemplars of Teaching Method', (Chicago: Rand McNally) pointed out that Plato had said that the market for second-, third-, and fourth-rate schooling would always be exploited by teachers who are themselves less than first-rate.

They went on to say that *'low-inspiration learning environments are made all the more possible and likely when teaching is formalized and methodized, because the school is attempting to go through the motions of educating, without invoking the knowledge and character of the first-rate teacher'*.

I fully concur with this observation. None of the boxes that can be ticked by a zealous Human Resources department can predict who your best instructors are likely to be. It is all about character, passion and zeal.

Appendix 1

Sixteen additional ideas for projects

- Design-and-build projects (numbered 1 to 7)
- Projects set around analysing a device (8 to 13)
- Projects requiring estimation, reasoning and creativity (14 to 16)

Design-and-build projects

Project 1: Maximum distance golf-ball thrower

A machine to project a golf ball a maximum possible distance, powered by the energy stored in a rope subjected to torsion. This photo shows one device built by students of the author. This machine achieved a throw distance of 49 m.

This student design might be less spectacular, but is still neat and workable. It is built at the level of technology expected from first-semester students for a project of this nature.

Project 2: Ramp flyer

Design and build a vehicle that must accelerate from rest on a flat floor, powered by the energy stored in rubber bands, then proceed up a ramp and launch itself as far as possible from the end of the ramp before returning to floor level.

The ramp would be provided by the department, and would have dimensions as follows:

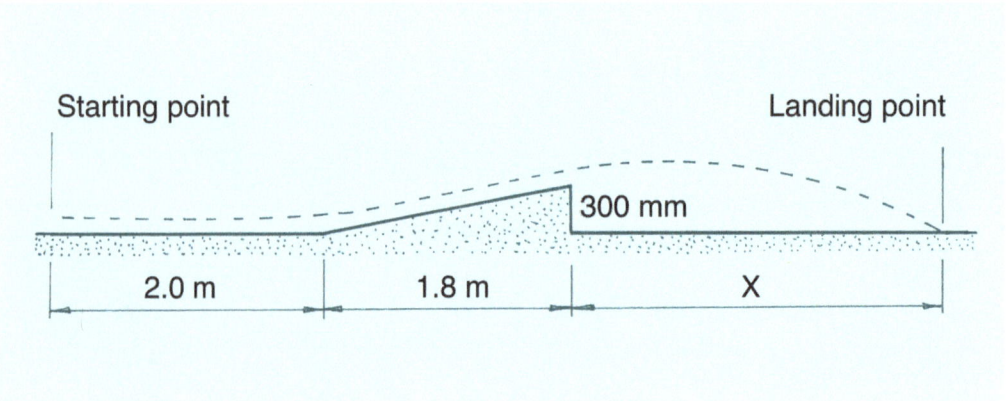

This machine's performance index will be the distance X from the end of the ramp to the landing point, divided by the weight of the vehicle.

There is no restriction on the materials you may use, except that you may not make use of a pre-manufactured item such as a plastic propeller or wheels or nails or screws. Wire and string are permitted. Every other part of the machine must be shaped by yourself. You may use a maximum of 12 rubber bands of a standard type provided by your instructor. The machine must be designed so that the rubber bands are easily removed, because on testing day you will be issued with a new set of the same type, that must be used in place of the ones you used for developing your machine.

The vehicle must have at least three road-wheels in contact with the floor when launched. The vehicle must remain in one piece for the duration of its ascent and flight. No part of it may become detached from the assembly. If it does so, whether by design or by accident, the shortest distance travelled by any part will count as the distance achieved.

The vehicle has to stay upright on landing. The distance achieved has to be greater than one metre, or your project will score zero.

In order to begin such a project, you need to determine, after suitable measurements:

a. How much energy can be stored in one of these rubber bands if stretched in a straight line.
b. How much energy can be stored if you use the rubber bands in torsion.

Describe what measurements you would need to make and how you would use the readings.

Then, determine:

c. How much energy will be available to drive your vehicle.
d. If your vehicle has mass **m**, how much of the original elastic energy imparted to the vehicle is still available as kinetic energy at the top of the ramp, and what your launch velocity is likely to be.
e. The take-off velocity needed to achieve a horizontal distance of 1.0 m, considering your vehicle as a projectile.

Explain what could cause the following problems, and how to overcome them:

- Wheel-spin on starting.
- A vehicle without propellers that continually veers to one side.
- A vehicle driven by a propeller, that continually veers to one side.
- Friction in the axle-bearings.

Explain how each of the following factors would influence the operation of such a ramp-flyer:

1. For a vehicle with driving wheels: given a particular diameter of driving wheel, would it be better to give the driving axle a large diameter or a small diameter?
2. Should the driving wheel/s have a larger or a smaller diameter?
3. How does the diameter of any non-driving wheel/s affect the performance?
4. Would it be better to have wheels with hard edges or with tyres?
5. Is it preferable to have the vehicle under power for a short time (a quick burst of energy) or a longer time (releasing energy slowly for the duration of the whole movement up the ramp)?
6. Describe, with the aid of a sketch, a way of converting a relatively large force, acting over a short distance, to a smaller force, acting over a longer distance. Some device like this might be necessary, considering that rubber bands supplied are originally 200 mm long, and can stretch to a maximum length of 350 mm.

Project 3: Standing long-jump machine

You need to build a machine that can launch itself off level ground (lawn on a sports field) and travel as far as possible in the air before landing. This would be a jumping machine, not a flying machine. It would need to be powered by energy that can be imparted by the operator, and stored as mechanical energy in one of three ways: either by torsion in a rope, or by the flexing of a wooden member, as in an archery bow, or by the stretching of rubber bands or springs.

The machine may *not* consist of two separate assemblies, one of which stays put while the other is flung through the air. The whole machine has to move as one, and no parts may come off during its operation.

The presenter may not throw the machine, or otherwise physically assist in its launch. The energy that was imparted by the muscular effort of the person presenting the machine, must be stored and then released by a trigger mechanism. This energy must be put into the machine in the presence of the adjudicators. It may not be stored prior to the testing day.

The maximum dimensions of the machine at any point in its operation may not exceed 1 m x 0.5 m x 0.5 m high. The machine has to stand unaided on the starting line. It must land on the side that was in contact with the ground at the time of launching.

Materials allowed: wood, particle board, bamboo, reed, cardboard, paper, metal fasteners, wire, string, rope, rubber bands and glue. The minimum distance that qualifies the machine for a successful jump is one metre.

The performance index = (distance achieved) ÷ (mass of machine)

This is essentially a brainstorming/design activity, making use of your knowledge of the principles of mechanics.

Questions:

1. What are the chief considerations that need to be taken into account when starting this design?
2. Name the principles of mechanics that are going to apply to the machine you would design.
3. Which materials would you favour?
4. Which method of storing energy do you feel is most effective for this particular challenge?
5. How would you try to ensure that your long-jump machine did not rotate or tumble sideways once it had taken off?

Project 4: Moon buggy

A 'moon buggy' vehicle that is powered by rubber bands, that has to approach and climb over a given small obstacle. The performance criterion is the distance travelled along the demarcated path *after* negotiating this obstacle.

Project 5: Pin-jointed bridge

This exercise was carried out in the days when we had few enough students to make it financially viable. A pin-jointed bridge had to be built of materials that were issued to the students. The materials consisted of sheet metal of a standard gauge, that they could cut into strips, and steel rods of a standard gauge, for pins. These bridges were tested to destruction on a purpose-built jig which the students had access to while building their bridges. The performance index was: (load at failure) ÷ (mass of bridge).

For details of this project and the testing jig, see Basic Engineering Mechanics Explained, Vol 1, p 159 - 161

Project 6: Accurate tennis-ball thrower

A machine powered by stretched rubber bands, that has to hurl a tennis ball accurately into a cardboard box such as those in which A4 paper reams are packaged, at a nominated distance from the front of the machine. Nine of these open boxes are taped together in a square, three boxes by three boxes. The central box is placed at the distance nominated by the assessor.

The instructor can nominate any distance between 2 m and 5 m from the front of the machine, in 200 mm intervals. These distances can be different for different competitors. The machine has to be calibrated for this range of distances, and has to have a trigger.

The performance criterion is the score from ten throws, with balls landing in the central box scoring 5 points, those landing in the similar-sized boxes attached to the central box being awarded 2 points, and those missing all the boxes scoring zero.

Project 7: A geared hoist

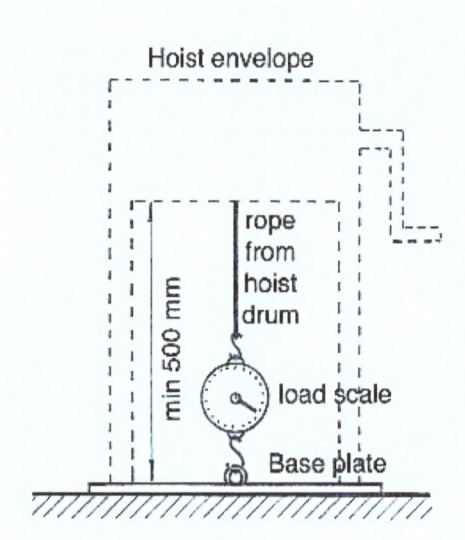

Aim: to build a geared hoist that can be used to exert an upward force against a spring scale hooked to a plate on the floor. All hoists will be progressively loaded until failure. The performance index will be:

P.I. = (max. load before failure ÷ mass of hoist).

Each group of three students has to build a geared hoist, consisting of a frame supporting two shafts, all made of wood. On the load shaft is to be mounted a drum and a gearwheel.

On the effort shaft is to be another gearwheel, meshing with the former one, and this shaft has to be turned by a hand-crank or wheel by one of the group members. On testing day this hoist has to be operated by one person only, without assistance of any kind, neither to turn the crank nor to steady the machine.

Wood is to be the predominant material. Bamboo is allowed, as are plywood, particle board, leather, rope, string, glue and nails, as long as these are shorter than 60 mm. No other metal parts may be used. All gearing is to be made either by cutouts in plywood or particle board, or in the manner of medieval gearing, with 'teeth' consisting of wooden pegs mounted in holes on wooden gearwheels.

The frame must be tall enough so that the lowest surface of the load drum is at least 500 mm above the floor, to leave space for the spring scale that will be supplied. This dimension is essential for the manner in which the machine will be tested.

Each machine will be placed on a testing jig, consisting of a steel plate on the floor, with dimensions 800 x 1200 mm. A hook is welded to the plate in its centre. The load rope will be attached to a spring scale that engages with this hook.

Size limitations: The entire machine, including protruding parts like a crank handle, must not exceed 1.2 m high, with a maximum floor plan of 0.8 m x 1.2 m. In order for the tension in the hoist rope to be measured, the machine has to stand on the steel plate which has those dimensions, so the legs of the machine must be suitably positioned. Load limitations: the machine must be able to sustain a load of at least 200 N. Failure to do so will result in a zero score.

Transgression of any of the rules results in a zero score.

The photo shows the working parts of one geared hoist built by a student team. The top of the spring scale can be seen at the end of the load rope. The steel plate on the floor was rusty, hence the brown colour.

b. Projects set around analysing a device

Project 8:   Leonardo's Dredger

Below is a reproduction of a drawing by Leonardo da Vinci, showing a design for a dredger. The dredger is built with two hulls, supporting a mechanism that is supposed to scoop mud off the bottom of a canal or lake, and deposit it into a barge for removal.

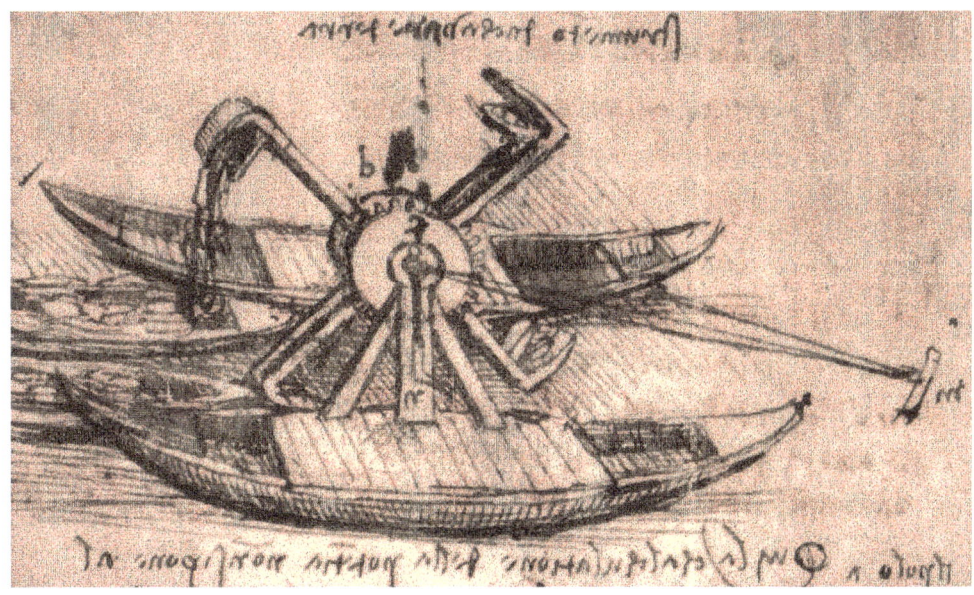

Some questions:

1. What is the intended power source?
2. Estimate the distance from the deck to the axis of rotation of the crank.
3. What is the function of the stake protruding from the water to the right of the vessel? Which principle of mechanics is operative here?
4. Has provision been made for dealing with mud at different depths? If not, propose a way of doing this.
5. Some gear teeth are visible. What are they supposed to engage with?
6. One scoop is shown depositing mud into the barge. However, when it has dropped its load and rotated further, it looks as if it will strike the barge. Suggest a modification that will prevent this from happening.
7. Suppose the scoops each contain 20 litres of mud. Estimate the radius of the scoop arms, and the torque required to turn the crank, if the entire rotating assembly is made of wood with the same density as oak. Assume the frictional torque in the bearings is 50 Nm.
8. Why are two barges shown behind the dredger?
9. Is this a viable design, given the power source available in the period in which da Vinci was alive?
10. Which features of Leonardo's sketch appear to be good ideas?

Project 9: Chariot

Research the designs of chariots as used in warfare, for instance, in ancient Egypt, or Rome. Which materials were used? What characteristics of such a chariot would you think are important?

Use your findings from the above research to design a chariot that students could build, which would be able to take part in a race, where pairs of students would take turns in being the 'horse' and the rider. The only part that is allowed to be made of metal would be the axle.

Which principles of mechanics would you need to be mindful of in your design? What are the features you consider essential for the safe and efficient operation of such a racing chariot? You don't need to build one, but your drawing and annotated labels should show that you have thought about the project intelligently, as if it were going to be built.Take into account the cost, which materials are suitable, and the level of manufacturing skills possessed by the members of your group.

Project 10: Ship disabler

In the harbour of Syracuse, in ancient Sicily, which was then part of the wider

Greek civilisation, written accounts assert that for the defence of the harbour, a mechanism was designed by Archimedes that could overturn any enemy ship that might enter the narrow harbour entrance. This mechanism was known as 'The claw of Archimedes' and also as a 'ship-shaker'. No diagrams of this mechanism have been found, but it was said to have been like a giant crane that dropped a grappling iron onto a ship and pulled it upwards, shaking the sailors and soldiers into the water. It is not confirmed whether such a mechanism was actually built. However, considering the technology of the ships and weaponry of that time, it would have been desirable to render an attacking ship useless by some means that could be controlled from the harbour.

Despite the spectacular nature of the reported device, it seems more practical to be able to take an enemy by surprise, by having the business end of the mechanism entirely underwater. Sketch a method of doing this, if the machinery is operated from a protected position adjacent to the harbour entrance channel. You may only suggest solutions that make use of the technology available in that era. Assume that power comes from teams of oxen pulling on ropes. The machinery has to be reusable.

Name the variables you would need to research to determine how much force would be needed to overturn (or otherwise render useless) a bireme of the time. Propose a way of converting the horizontal force exerted by a team of oxen to the type of force needed. Show your reasoning and calculations for how many oxen might have been required.

Project 11:   A working model mangonel

This drawing of a small working model mangonel shows a design built by students. The objective of this device was to hurl a golf-ball as far as possible over level ground. It is built mainly of pine wood, and complies with the dimensional constraint that the machine had to fit into a cube that is half a metre on each side.

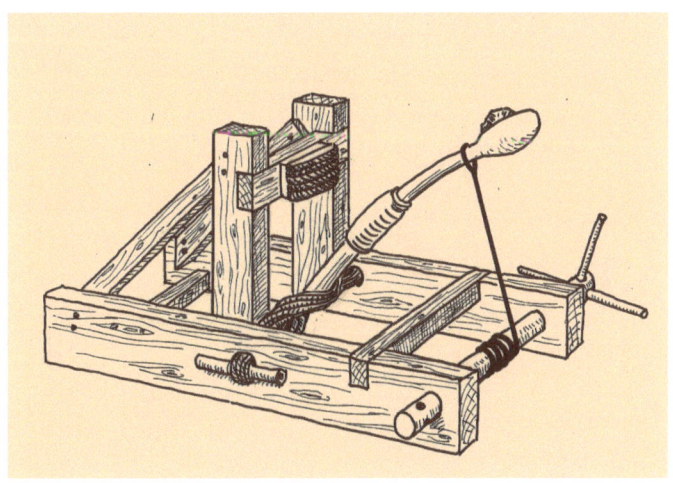

1. Identify which other materials (besides pinewood) might also have been used.
2. Identify the function of each part of the machine.
3. In how many ways is energy stored in this machine, in the position shown?
4. Explain the adjustments that would have to be made to this design to optimise the range of the projectile.
5. This design does not show a trigger that can be used to fire the mangonel. It looks as though the string would have to be cut to release the throwing arm. Can you suggest a trigger mechanism that would release the string without cutting it?
6. What would be the effect on this machine of the recoil when it is fired? Would firing a shot drive the machine backwards or forwards? Why?
7. List all the considerations you can think of that might affect the efficiency of the machine, and what could be done to optimise its efficiency.

Project 12 A student-built gravity-powered vehicle

This diagram shows a wooden vehicle built by students for a design-and-build project. The vehicle is powered by the gravitational potential energy released by a descending mass-piece. The intention was for the vehicle to cover the greatest possible distance on a flat, level floor.

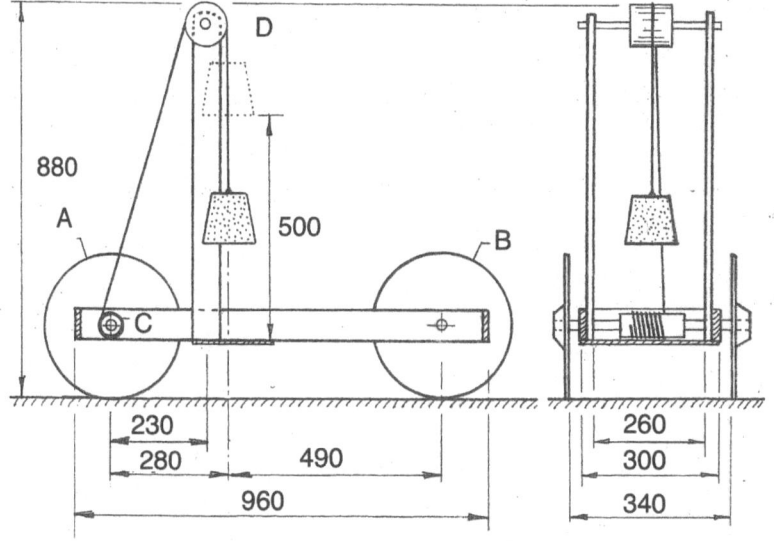

The string attached to the mass-piece causes the rear axle to turn, thus propelling the vehicle forwards.

The metal mass-piece is released near the top of the tower when the vehicle is at rest, and is allowed to descend 500 mm. At this point, the string comes off the driving axle. The mass-piece lands on a platform which is part of the vehicle, and travels with it until the vehicle again comes to rest.

The rectangular frame of the vehicle (including the tower) is made of pine wood, of cross-section 70 x 10 mm. The wheels and platform are of 3 mm hardboard. All axles are made of 10 mm diameter wooden dowels. The initial diameter of the drum on axle **C** is 60 mm.

The masses of the respective components are, in [kg]:
Lower frame (chassis) 0.45
Tower plus sheave D and shaft 0.30
Rear wheels plus axle.. 0.30
Front wheels plus axle 0.25
Descending mass-piece 2.0

Answer the following questions with reasonable accuracy, giving explanations where required, and stating assumptions where these have had to be made:

1. What amount of energy is available to drive the vehicle?
2. Determine the normal reaction of the floor on the driving wheels.
3. Determine the initial torque available at the driving axle.
4. What is the magnitude of the force available at the perimeter of the driving wheels, to push the vehicle forwards?
5. If the coefficient of friction (μ) between the wheel perimeter and the floor is 0.4, will the driving wheels spin on starting?
6. Assuming linear acceleration and deceleration, the following two graphs could represent the motion of this vehicle:

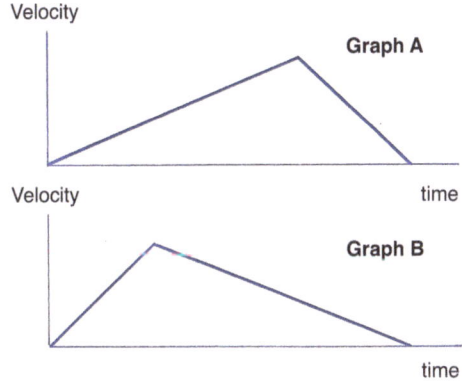

Would either of these two graphs be associated with a more energy-efficient vehicle? Why?

7. If a vehicle of this type travelled a total distance of 25 m, taking 20 seconds from start to finish, what would have been its maximum velocity?
8. If the total distance travelled was 25 m, how much work was done against friction?
9. If the total distance travelled was 30 m, how much work was done against friction?
10. Would the vehicle travel further if (a) the mass-piece remained on it, or (b) the mass-piece was allowed to fall off it, after descending 500 mm? Why?
11. Name the principle of mechanics that is applicable to your explanation for question 10.
12. What would be the effect on the vehicle's performance, if the wheels were not perfectly circular?
13. What effect would be noticed if the plane of the wheels was not perfectly aligned with the direction of travel?
14. If all other dimensions remained the same, but diameter **C** changed as follows, complete the following table:

Diameter 'C' [mm]	60	30	16	10
Distance in metres covered by the vehicle while still under power (before the string comes off the driven axle)				
Initial torque available when the mass-piece is released [in units of Nm]				

15. Is there any advantage to be gained by making the diameter of the front wheels smaller than that of the rear wheels? Explain.
16. What is the effect of increasing the diameter of sheave **D**?
17. Do you think it is desirable to do something to reduce the air resistance of this vehicle, in order to make more efficient use of the available energy? What could be done? How much of a difference would it make to the performance?
18. Name two things you could do, aimed at reducing friction losses in this vehicle.
19. Consider using a long rubber band as a tyre on the driving wheels: is this a good idea, or not? Why?

Project 13   Human-powered water craft

Humans have made rafts and boats move on water for millennia. There are a number of ways in which the force exerted by a person on a floating object can be made to achieve thrust. Traditional modes of providing thrust are paddles and oars. More recent developments include pedal-powered submerged propellers, paddle-wheels, and air propellers.

Illustrated here is an alternative design that makes use of a simple flap valve as found in a lift pump. There are two cylinders, one inside the other. The outer one is fixed to the craft. The operative idea is that, on the pull stroke, the flap valve closes, so the water in the inner cylinder is pushed backwards, providing thrust. On the forward stroke, more water enters the inner cylinder.

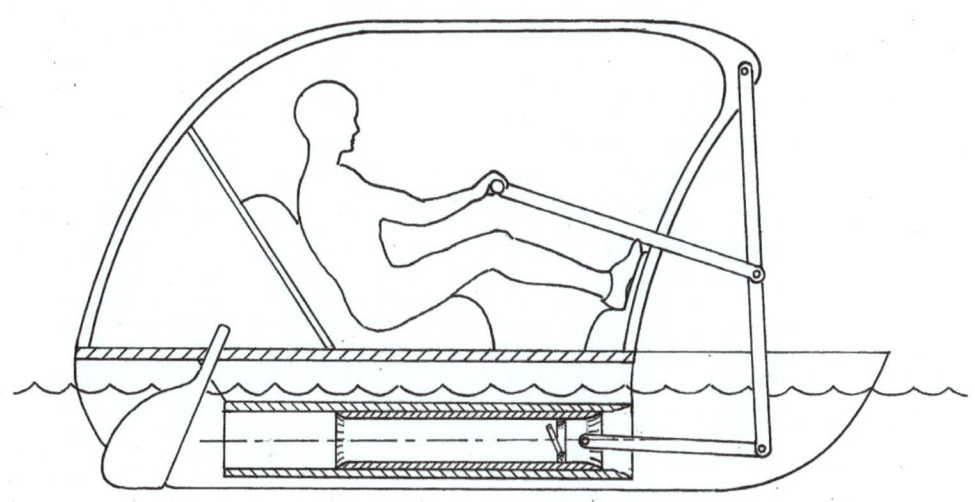

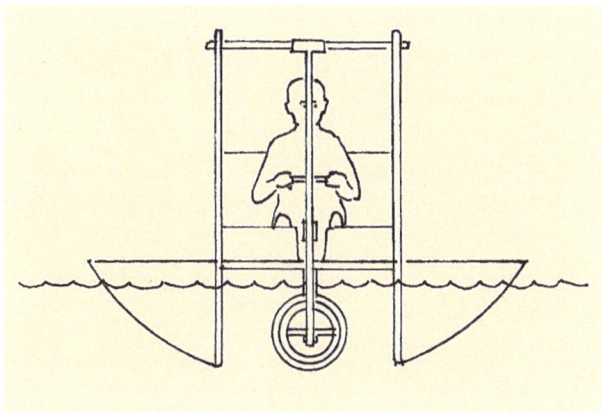

The craft is double-hulled. A rudder is shown, but the linkage that controls it is not shown, as the means of steering is not relevant to the questions about propulsion below:

1. Would this concept for achieving thrust actually work?
2. If it would work, which principle or law of mechanics explains *why* it works?
3. On the forward stroke, is there any resistance offered by the water? If so, where? What could be done to minimise this resistance?
4. Does the outer cylinder have a function besides guiding the inner one?
5. What considerations affect the choice of how deeply submerged the cylinder needs to be? Explain.

6. What considerations affect the dimensions of the cylinders, in both length and diameter?
7. What material would be suitable for the walls of the cylinders?
8. What material would you suggest for the frame and the levers? Why?
9. Is a watertight seal between the two cylinders necessary? Why, or why not?
10. Which would suit human ergonomics better, to have a long, slow stroke, or a short, punchy one?
11. How high above the waterline would you estimate the centre of gravity of this craft to be, with a rider on board?
12. Would such a craft appear to be stable? Explain why or why not, in relation to the shift in position of the centre of gravity and the centres of buoyancy of the two hulls, when the craft tilts slightly.

c. Projects needing estimation, reasoning and creativity

Project 14   The Lighthouse Build

In this exercise, you are not given all the data you require. You have to make reasonable estimates of certain values, in order to tackle the problem. Your group's solution will be assessed on the clarity with which you explain your reasoning. The numerical value of the answers is less important.

A lighthouse has to be built on a steep hill ending in a cliff overlooking the sea. The site is on an island, in an area where the average number of rainy days per month is eight. Portable engines are not available, and the only source of power is human effort. An electricity supply will be led to the site only after the build is complete.

The lighthouse is going to be roughly in the shape of a truncated cone, 10 m high, 7 m in diameter at the base, and 3 m in diameter at the top, with wall thickness 0.5 m. An internal stairway will need to be built, spiralling up the walls.

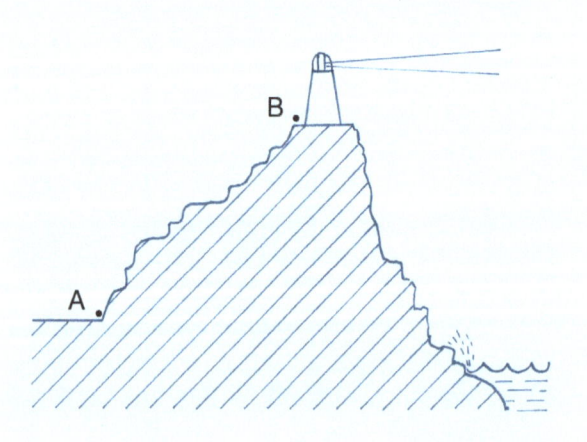

a. If it normally takes 420 bricks to build a solid cubic metre of structure, how many bricks will be needed?
b. When *not* carrying a load, a man can climb from **A** to **B** (an elevation of 40 m) in ten minutes, and come down again in 5 minutes. The path is rough and rocky. If the average worker on the job weighs 80 kg, at what rate does he expend energy while doing this climb? Assume that one of these workers can keep up this rate of expenditure of energy for an hour before needing a five minute rest.
c. These bricks would have an average mass of 2.7 kg, and there are 15 men available. To get the bricks to the top, consider three possible scenarios:

- Carrying the bricks by hand from **A** to **B**
- Using a human chain to pass the bricks up the steep hillside
- Rigging up a lift powered by human effort, to move the bricks

The questions that follow lead you to decide which one of these scenarios would seem to have the most attractions, from the point of view of energy expenditure:

d. If the bricks have to be carried *by hand* from **A** to **B**, and a working day is 8 hours, *how many working days* will it take to get all the required bricks to the building site?
e. If a human chain is used, with each man at his station relaying bricks up to the next man, would that be more effective or less effective than carrying them up from the bottom? Explain.
f. What else will need to be carried up the hill to the site? And how long would that take?
g. If you decided that, instead of carrying the building materials to the site, you would rig up a human-powered lift system, sketch your best idea for this lift, and identify which materials would be needed to build the lift. Besides some sheaves that can be obtained, the only materials available are timber, rope, bolts, steel shafts on which the sheaves can run, and fencing wire. What hand-powered tools would you need to build this lift?
h. Explain the way this lift system would have to be anchored, in order to counter the downhill pull exerted by the weight of the lift and the materials that it carries.
i. Give a reasoned estimate of the mechanical advantage of this lift.
j. How long would it take to build and commission this lift?
k. Would it be worthwhile building the lift, or more efficient to proceed with one of the other two scenarios?
l. If there are three bricklayers, each able to lay 600 bricks in a day, will the supply of bricks outpace the rate at which they can be used up? Bear in mind there is virtually no room at the top of the hill to stack materials.
m. If you were the contractor for the job, what would be your overall time estimate for the whole job, to advise the person commissioning this work?

Project 15 The tennis ball thrower

Suppose that you have to design a machine to throw a tennis-ball the greatest possible distance over a level sports field, using energy stored in the machine, provided by the muscular effort of the person operating the machine. The only materials allowed are wood, string and rope made of natural fibres, canvas, screws, nails, rubber bands of any size, and a lubricant. The machine may not occupy more than a space with the dimensions 1m × 1m × 1m, at any stage in its operation. The machine must be operated by only one person at a time. At the time of operation, no persons other than the operator may be touching the machine. Now, answer the following questions:

1. Choose the method of propulsion: Which would be better: storing the energy for the throw in a twisted rope (as in Roman trebuchets), or in a rubber strip under tension, as in a slingshot, or some other method, like the flexing of a strip of wood, as in an archery bow?
2. What are the strength considerations for the materials you intend to use?
3. What provisions should you make to absorb recoil?
4. What is the ideal angle at which to launch the projectile?
5. Is it more effective to strike the ball suddenly, or to accelerate it gradually? Why? Would your answer be the same if the ball were made of solid wood?
6. Which parts of the machine need to be heavy, and which parts light? Why?
7. What is the best way to lubricate any moving parts, considering the materials they are made of?

Project 16 The raft

Suppose you are one of two explorers who find yourselves stranded with your boat smashed beyond repair. The engine has sunk in a deep part of the river. A raft must be built to carry you and your gear (pared down to 50 kg) downstream along the river for a long distance, possibly 200 km. The conditions ahead are unknown. There might be rapids. The raft might need to be carried or taken apart occasionally to get it through narrow gaps. You have salvaged from your smashed boat twenty empty 20-litre cylindrical fuel cans which you could use for flotation. Assume that these cans are airtight when the lids are fastened on. The only other materials available are freshly cut local timber from the forest, and 100 m of 6 mm diameter nylon cord. Since the fresh-cut wood will be full of sap, you can't bank on it having a density low enough to aid significantly in flotation. Assume the density of the wood is similar to that of oak. You have an axe, a spade and a crosscut saw, and each of you has a sheath knife. Assume you have no other tools, cord or fasteners.

1. Determine the likely weight of one fuel can, and the buoyancy that it could provide.
2. What dimension of tree-wood would you look for? Logs or saplings?
3. Sketch a proposed layout for the raft, in plan and in section.
4. What is the likely weight of the raft plus people plus load?
5. How would you ensure that the surface of the raft is above water while in normal slow-water operation?
6. How many fuel cans would be needed for flotation?
7. How would you ensure that the cans stayed in place?
8. Would the raft best be flexible or rigid? Explain.
9. How would you steer the raft?
10. How would you propel the raft?
11. In rough-water conditions, is it better to have a light raft, or a heavy one? Why?

Appendix 2

Eleven additional examples of calculation exercises developed by this author

These are extracted from this author's book 'Revision Exercises in Basic Engineering Mechanics'. More than 120 others are to be found in that book, and a futher 420 in the three volumes of 'Basic Engineering Mechanics Explained'.

Exercise A2.1

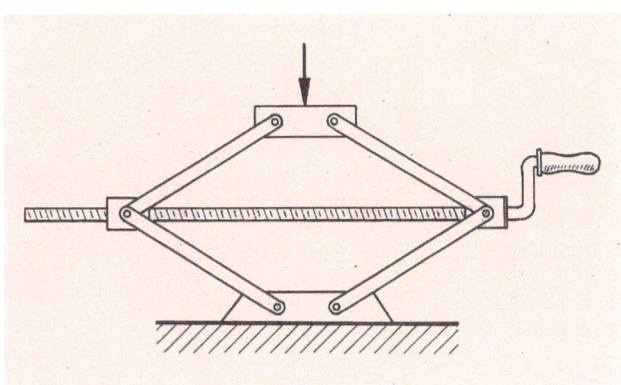

A motor car scissors jack is used to raise one side of a small car to change a wheel. The effective mass to be raised is 560 kg. The height by which it must be raised is 145 mm. Assume the efficiency of the jack to be 42%.

Draw an energy accounting diagram for this operation of the jack, to scale. Ignore the mass of the parts of the jack that need to be raised.

Using the diagram, determine:

- The work done by the person operating the jack. [1897 J]
- The amount of work done against friction. [1100 J]
- The time it would take to perform this operation, if the person could perform mechanical work at a rate of 75 watts. [about 26 sec]

Exercise A2.2

The diagram shows one side of a small pin-jointed bridge in the form of a Warren truss. The other side of the bridge is identical and parallel to this one. Assume that cross-bracing keeps the two sides stable in a vertical plane.

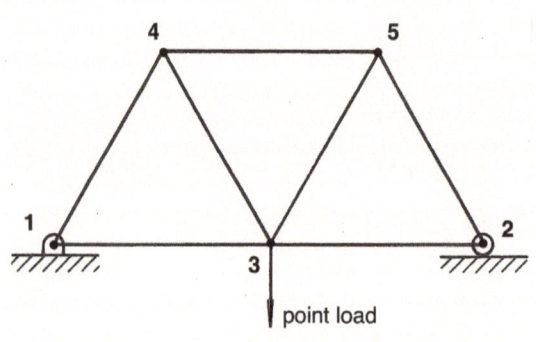

All the bridge members are 4 m long, and the bridge weighs 14 kN. A point load of 26 kN is suspended from node 3.

Analyse one side of the bridge only, to determine the tensions in each of the members shown.

[members **1- 3** and **3 - 2** are in tension: 5.774 kN;
members **1 - 4** and **2 - 5** are in compression: 11.55 kN;
members **3 - 4** and **3 - 5** are in tension: 11.55 kN;
member   **4 - 5** is in compression: 11.55kN]

Exercise A2.3

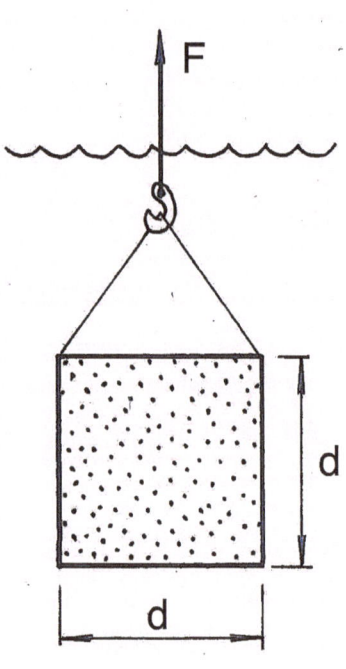

A crane has to pull a concrete cube (density 2400 kg/m³) out of a freshwater lake. The cube sides have dimension **d**. While the cube is being raised, but still submerged, the tension in the crane rope, F = 275 kN.

Determine:

- The value of **d** in metres. [2.715 m]

- The tension in the crane rope when the cube is out of the water. Consider the density of air to be 1.29 kg/m³. [471.2 kN]

Exercise A2.4

A train of three linked railway wagons, each of mass 20 tonnes, is freewheeling on a level stretch of track, at 20 m/s. Each of the wagons experiences a rolling

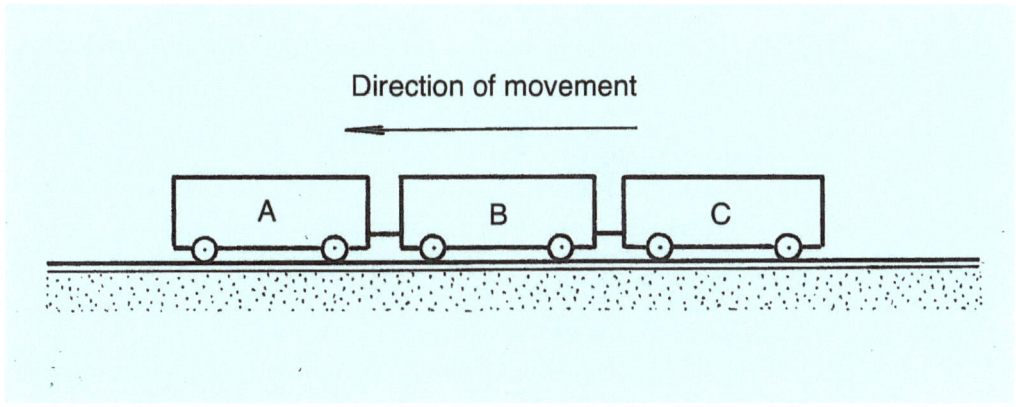

resistance of 3 kN and air drag of 2 kN at this speed. The emergency brake on the rear wagon is suddenly applied, providing a braking force of 100 kN.

Determine:

- The value of the deceleration of the train [1.75 m/s²],
- The tension in the link between the rear and middle wagons [70 kN],
- The tension in the link between the front and middle wagons [40 kN]
- The distance in which the train will be brought to rest, if the air resistance remained constant. [114.3 m], and
- Whether the braking distance would be greater or less than this value, if the variability of air resistance with velocity is taken into account.

Exercise A2.5

A hand-cranked winding drum is used to raise a weight of 300 N slowly through a height of 4 m. The wire rope passes over a fixed, solid steel cylinder of diameter 120 mm. The winch drum diameter is 200 mm and the crank radius is 350 mm.

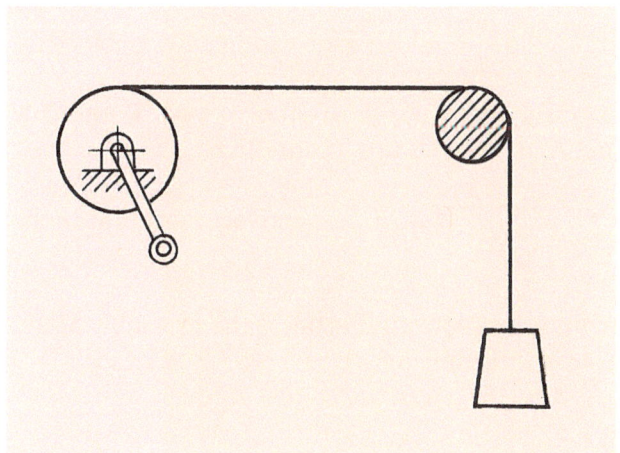

When a rope passes around a solid fixed cylinder, the ratio between the tensions in the tight side and the slack side of the rope is given by the equation:

$$T/S = e^{\mu\theta}$$

where e is the natural number 2.71828, also known as Euler's number, μ is the value of the coefficient of friction between the rope and the cylinder, and θ is the angle of contact between the rope and the cylinder, expressed in radians.

If $\mu = 0.4$ in this case, determine:

The tension in the wire rope going onto the winding drum [562.3 N],
The work done by the person turning the crank in raising the load through this height [2249J],
The percentage of this work that is useful work [53.35%],
How long this process would take if the operator was capable of a power output of 20% of one horsepower [15 seconds], and
The tangential force the operator would have to exert on the crank to hold the load steady, to just prevent it from slipping downwards. [45.73 N]

Exercise A2.6

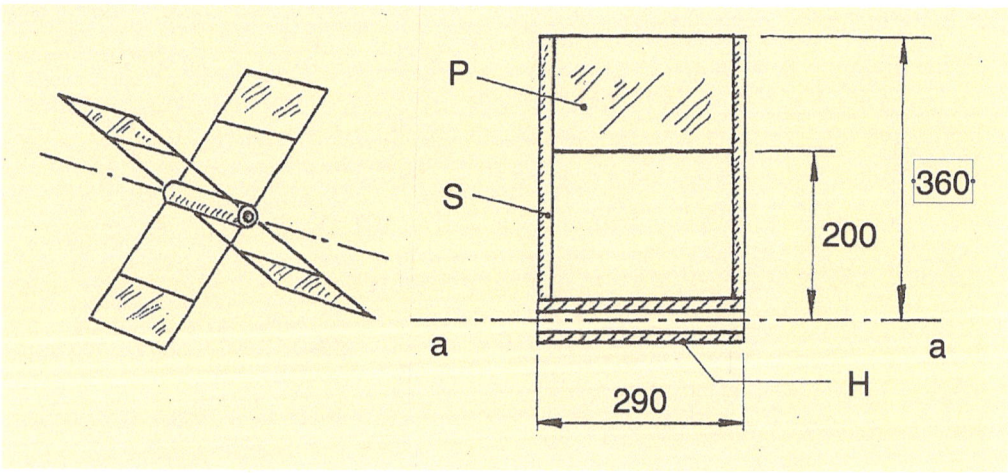

A paddle-wheel consists of a central hub **H**, to which four paddles **P** are joined by spokes **S**. All parts are made of brass, density 8400 kg/m³.
Hub **H** is of hollow circular section, OD 60 mm and ID 30 mm. Spokes **S** are made of solid round rod, diameter 20 mm. The paddles are 5 mm thick.

Determine the mass of **H**, one part **S** and one part **P**, respectively, accurate to the nearest gram. [5.166 kg; 871 g; 1.680 kg]

Using these values, determine the mass moment of inertia of the assembly about axis **a-a**, as accurately as possible, to four significant figures. [0.8723 kg.m²]

Exercise A2.7

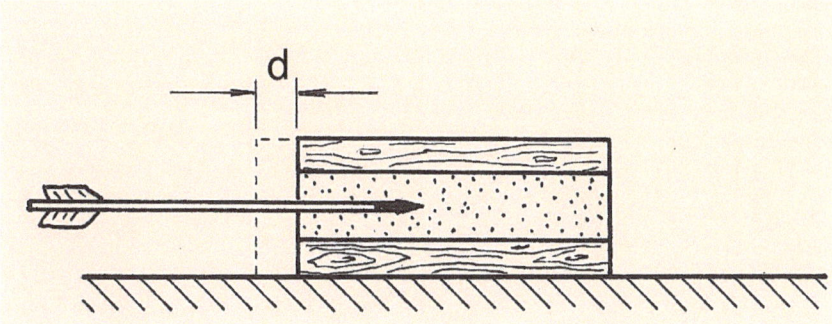

An arrow of mass 60 g is fired into a target that consists of a block of expanded polystyrene to which is glued two wooden slabs. This target rests on a flat horizontal table where the coefficient of kinetic friction between the wood and the table is 0.4. On impact, the arrow becomes imbedded in the polystyrene, and the target is found to move a distance d = 10 mm.

Determine:

- The amount of work done against friction with the table while the target is sliding. [0.2934 J]
- The velocity of the target immediately after impact, if it is assumed that 80% of the arrow's original energy before impact is lost on impact. [0.6245 m/s]
- The velocity of the arrow on impact. [104.7 m/s]

Exercise A2.8

A WW2 fighter plane sustains damage to its three-bladed propeller in a dogfight. A small piece gets shot off the end of one of the propeller blades. The mass of this missing piece is 3 kg, and its centre of mass was d = 1.45 m from the axis of rotation of the propeller shaft.

Given that the mass of the intact propeller was 168 kg, determine:

- The distance that the centre of mass of the propeller has shifted due to this damage, [26.36 mm] and

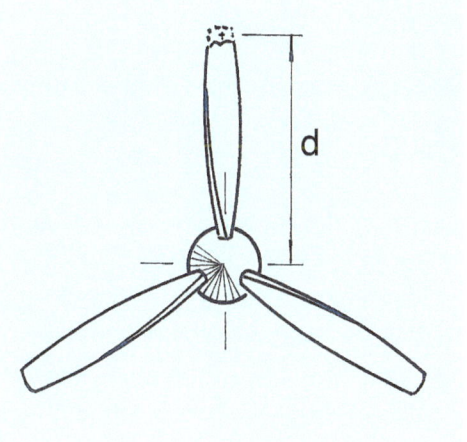

- The consequent out-of-balance force on the propeller shaft bearings at a rotational speed of 1320 r/min. [83.12 kN]

Exercise A2.9

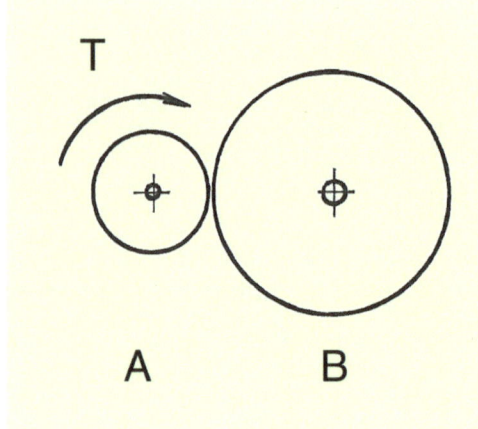

Two large gearwheels, **A** and **B**, mesh together. Consider them to be solid discs for the purposes of determining their respective rotational inertias.

Gear **A** has diameter 400 mm and mass 20 kg. Gear **B** has diameter 800 mm and is made of the same material as gear **A**, with the same thickness.

The gearwheels are initially at rest, when a clockwise torque of 40 Nm is applied to gear **A**, and is steadily maintained. If the frictional torque on the gear shafts is assumed to be negligible, determine:

The time it will take for gear **B** to reach an angular velocity of 100 rad/s. [20 seconds] and

The tangential force that the teeth of the gearwheels exert on one another during that time. [180 N]

The effective rotational inertia of this gear train, from the point of view of applying a torque to gear **A**. [4.0 kg.m²]

Exercise A2.10

This cast-iron flywheel has a complex shape, making it difficult to establish its mass moment of inertia by directly calculating the way the mass is distributed.

However, the value of the MMI can be determined by allowing it to swing with small oscillations in the plane of the wheel, suspended from a knife edge as shown, and timing the oscillations.

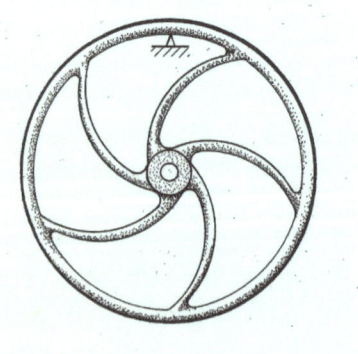

If the mass of the flywheel is 242 kg, the knife edge is 465 mm from the centre of the wheel, and the time taken for 10 oscillations is 19.24 seconds, determine the MMI of this wheel about its central rotation axis. [51.18 kg.m^2]

Exercise A2.11

The back wheel of a stationary bicycle is replaced with a flywheel fixed to a sprocket whose effective radius is 25 mm. This flywheel is a solid disc of stone, density 2400 kg/m^3, diameter 600 mm, and thickness 50 mm.

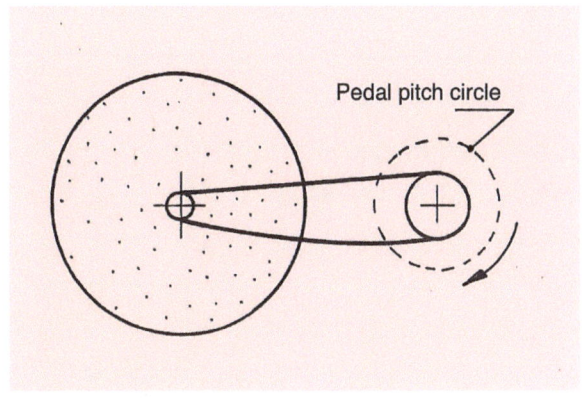

The rider can exert a pedalling torque of 15 N.m, and sustains this torque to bring the flywheel from rest to a speed of 3 rev/s.

The front sprocket's effective radius is 75 mm. The diameter of the pedal pitch circle is 320 mm. Consider as negligible: the rotational inertia of the pedals and front sprocket, and all frictional resistance. Determine:

The mass moment of inertia of the flywheel. [3.181 kg.m^2]
The change in angular momentum of the flywheel during this period of acceleration. [59.96 kg.m^2/s]
The time it takes to reach this speed. [12 seconds]
The tension in the upper part of the drive chain. [200 N] and
The average rate of expenditure of energy provided by the rider over this period. [47.09 W]

Appendix 3

Typical questions that can be used in lab tests

Lab Test Question 1

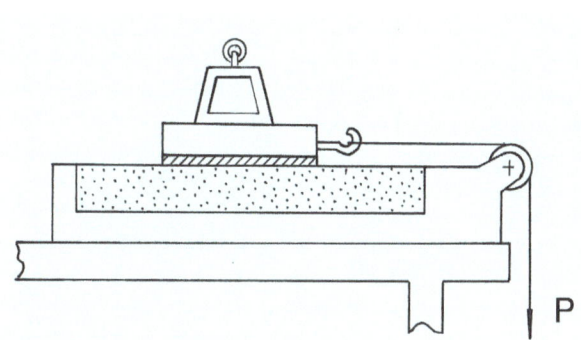

In an experiment to determine the coefficient of friction, µ, for rubber on concrete, a block with a rubber facing is used. This rests on a horizontal plane with a concrete facing.

A varying amount of mass is placed on the block. For each value of mass, force **P** is gradually increased from zero until the block just begins to move. At this point the value of **P** is recorded.

The combined weight of the block and the extra masses placed on top of it constitutes the normal compressive force **N**, measured in newtons.

Suppose the following readings are obtained:

N, measured in newtons	P, measured in newtons
8.0	6.5
14.0	10.5
18.0	15.8
24.0	19.5
32.0	25.0

Plot these readings on a graph that you draw on the graph paper provided, and from the graph, deduce the value of the coefficient of friction, rounded off to two decimal places. [answer: 0.80]

Though not stated in the test paper, marks for the graph and the processing of information resulting from the graph would be allocated for the following criteria. These are all criteria that the students would have been informed about in their lab instructions:

a. Appropriate heading and sub-heading, if applicable.
b. Well-chosen scale.
c. Proper labelling of axes.
d. Placement of graph on the page.
e. Conventional plotting of points and deduction of the 'best-fit line'.
f. Showing clearly how the value of the coefficient of friction is deduced from the graph.
g. Accurate value of the coefficient of friction obtained.

Lab Test Question 2

Concerning the experiments you did to investigate the relation between F_{max} and N, and to determine the coefficient of friction:

a. Name three factors that could negatively affect the accuracy of the readings taken during this experiment.
b. Which variable is the most difficult to establish an accurate value for?
c. Why is it necessary to take several readings of F_{max} at each value of N?
d. Why does it lead to a more reliable result if we plot all the readings on a graph and find a 'best fit' line for the graph, rather than processing the values obtained for one reading only?

Lab Test Question 3: a true/false test

a	All stopwatch readings could be regarded as inaccurate, on account of the reaction time of the person using the stopwatch.	T	F	?
b	If two values are obtained for a variable, say A and B, where A is the larger of the two, then the percentage difference between the two is given by $((A - B)/B) \times 100$	T	F	?

c	The best-fit line on a graph derived from experimental readings should pass through all the plotted points.	T	F	?
c	If six people take a stopwatch reading of the same event, and one of these readings looks significantly different from the others, that reading should be disregarded.	T	F	?
e	On the centrifugal force apparatus, if the levers click over at 45 r/min while the rotating arms are accelerating, they would click back at 45 r/min when it is subsequently slowing down.	T	F	?

Lab Test Question 4

These questions count 5 marks each, and require short written explanations:

a. In the experiment you did on the centrifugal force apparatus, what were you supposed to be comparing?
b. Describe briefly the method of an experiment to determine the value of the gravitational acceleration, 'g', using a simple pendulum.
c. Sketch and label the graph you obtained in the experiment to verify the equation $T/S = e^{\mu\theta}$.
d. Describe what is likely to happen in the experiment with the spinning chair, if you used weights that were twice the mass of those provided.

Lab Test Question 5

An experiment was done to investigate the vertical oscillating behaviour of a spring-mass system. The aim was to attempt to verify the equation that relates the period of oscillations to the values of the suspended mass and the spring constant.

The spring constant was found to be 200 N/m. Using stopwatches, six students each timed 10 complete oscillations. Their readings, in seconds, were as follows:

Suspended Mass, [kg]	student A	student B	student C	student D	student E	student F
4.0	8.8	8.9	8.8	9.7	8.8	8.9
5.0	9.9	9.8	10.0	10.1	9.9	9.8
6.0	10.9	10.8	10.9	10.9	10.7	11.0

- Something went wrong with one of the readings. Identify which one, and describe what probably happened to cause this anomaly.
- Process this data to show whether or not the results verify the equation.

Lab Test Question 6

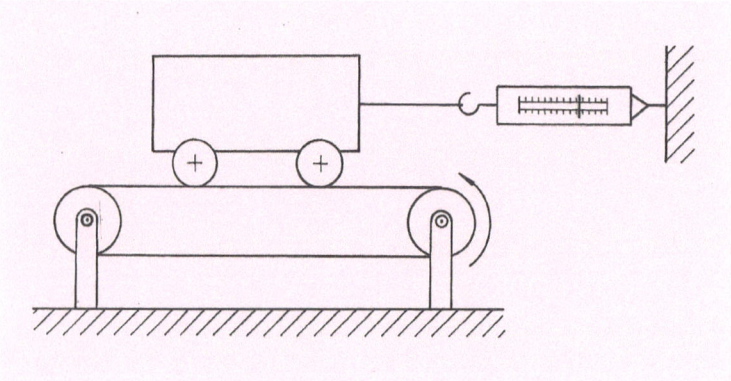

This diagram shows an apparatus that is proposed for an experiment to determine whether the rolling resistance of a small wagon varies with speed and/or load. The speed of the conveyor belt 'treadmill' can be controlled and measured. Additional masses can be placed in the wagon.

Describe, in clear steps, *only the **method*** of an experiment to determine whether rolling resistance varies with load.

Style: Write this description as if you are reporting an actual experiment that you have carried out. Do not write it in the style of a set of instructions. If you do, you will score zero for this question.

Lab Test Question 7

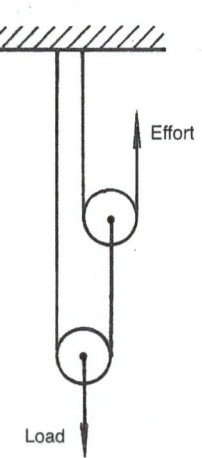

The diagram shows a system of sheaves and tackle used as a simple lifting machine. Describe in your own words: When investigating the properties of this machine in the lab:

a. How do you physically measure the effort? Name the instruments and all the precautions necessary to obtain reliable readings.

b. How do you determine the velocity ratio?

Lab test Question 8

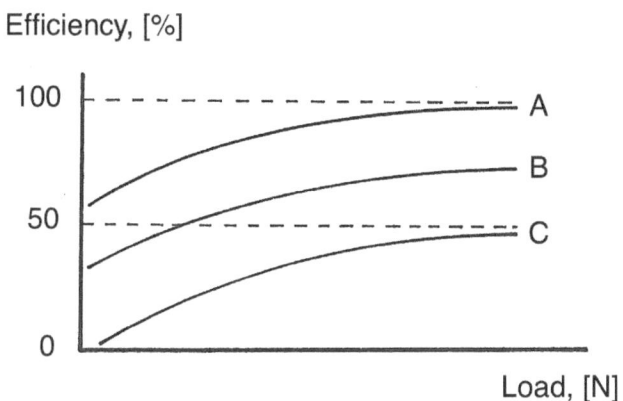

The graph shows three possible curves of efficiency vs. load for simple lifting machines.

Identify which of the curves would apply respectively to each of the following machines, and give your reasoning:

- A wheel and differential axle.
- A car jack based on turning a nut on a screw thread.
- A block and tackle used for raising loads.

Lab Test Question 9

Write a description of the **method** of the experiment you did to investigate the relationship between the tight side tension and the slack side tension of a cord passing around a fixed circular drum.

Do not write it in the style of a set of instructions. If you do, you will score zero for this question.

The description should contain a clear diagram of the apparatus. The whole description, including the diagram, may only occupy one side of one page. If your answer spills over to a second page, the rest of it will not be marked. Plan it carefully, in rough, before you write it out.

Lab Test Question 10

Suppose you are trying to determine the braking effort on a shaft, provided by a particular type of mechanical brake: consider two possibilities:

1. The shaft is stationary, with a known torque applied to it attempting to rotate the shaft, while the brake is attempting to prevent rotation, and

2. The shaft is rotating slowly, with a known torque causing the rotation, while the brake is attempting to resist rotation.

Which of the two circumstances would result in a more reliable value for the braking torque, and why?

Lab Test Question 11

Some measurements are made to determine the value of the spring constant of a given spring. On a subsequent occasion, another investigator determines this value, to check whether the same result is achieved.

The values obtained are, respectively: 1276 N/m and 1250 N/m.

- If it is not known which of the two results is more reliable, would the percentage difference between these two figures best be described as 1.72% or 1.76%? Explain.

- If the measuring equipment used was the same on both occasions, and if each of the results was obtained by taking at least eight different measurements of force and extension to use in plotting a graph, what could account for the different results on the two occasions?

Appendix 4

Student Feedback Survey Form

If one is to attempt to bring about improvements to a course, it makes sense to have a means of deciding how well those 'improvements' are being perceived by the students.

For this purpose, the author has developed a student feedback form that is based on his research and experience in the field of gathering feedback from students. This form is presented here, and may be found useful by instructors.

The first table seeks to find out about the general situation of the respondent in relation to coping with the course. This is important because feedback should not be admissable unless you know how well qualified the respondent is to provide that feedback.

The second table is what I call a list-format survey. All the course features deemed important by the instructor are named in a list. Respondents indicate their satisfaction with each item on the list. Doing that serves to remind them of what they experienced on the course. They can then identify issues that they might wish to raise. If there are any issues, they have to be described in the respondent's own words, otherwise the instructor cannot properly assess whether a criticism is legitimate and needs to lead to change.

It takes students about 15 minutes to complete the survey. Results are easy to process. It is good practice to inform the students of the main findings of the survey, to assure them that their instructor takes their feedback seriously.

Note: You do NOT need to put your name on this form!

Course: ....

Instructor..Date...........................

Make a tick mark (✓) in the appropriate box	0 - 20	20 - 40	40 - 60	60 - 80	80 - 100
What percentage of all the learning activities provided on this course have you attended up to now?					
What percentage of all the issued homework have you done up to now?					
In which range is your highest mark for any assessed event up to now?					
In which range is your lowest mark for any assessed event up to now?					
What score as a percentage do you realistically expect to get in the final assessment?					
Number of times you have consulted a tutor	0	1	2	3	> 3
Number of times you have consulted the instructor, whether in class or by appointment	0	1	2	3	> 3

For each of the questions or statements about the course below, **circle** one of the following symbols:

★★ = This is great, excellent, outstanding
★ = This is quite good, I am happy about it.
OK = Acceptable, no comment, nothing to say
− = Not really satisfactory
− − = I have a problem with this

If any item in the list below did not apply on this course, you do not need to respond to it.

1	How well do you usually follow the explanations given by your instructor?	★★	★	OK	−	− −
2	How well do you understand the principles you are supposed to be learning about?	★★	★	OK	−	− −
3	How well have your previous studies (before you took this course) prepared you for work at this level?	★★	★	OK	−	− −

4	Are you learning enough that is new to you, i.e. not going over old stuff?	★★	★	OK	–	– –
5	How interesting do you find this subject?	★★	★	OK	–	– –
6	How well does your instructor appear to know the subject?	★★	★	OK	–	– –
7	The pace at which he or she explains during lectures	★★	★	OK	–	– –
8	The variety of learning activities used in this course	★★	★	OK	–	– –
9	The way your instructor maintains discipline in class	★★	★	OK	–	– –
10	How well-prepared for class your instructor seems	★★	★	OK	–	– –
11	Your instructor's attitude towards the subject	★★	★	OK	–	– –
12	Your instructor's attitude toward students in general	★★	★	OK	–	– –
13	How well do you understand the kind of language your instructor uses?	★★	★	OK	–	– –
14	Your instructor's availability when you have questions to ask	★★	★	OK	–	– –
15	How approachable is your instructor when you want to consult him/her?	★★	★	OK	–	– –
16	Your instructor's general liveliness and sense of humour in class	★★	★	OK	–	– –
17	Your instructor's professionalism: punctuality, fairness and manners	★★	★	OK	–	– –
18	The breadth of knowledge displayed by your instructor around the subject	★★	★	OK	–	– –
19	The course materials and handouts supplied by your instructor	★★	★	OK	–	– –
20	The course arrangements: is it always clear what is expected of you?	★★	★	OK	–	– –
21	The lecture and tutorial venues	★★	★	OK	–	– –
22	The lab instruction booklet	★★	★	OK	–	– –
23	The lab venues and equipment	★★	★	OK	–	– –
24	The availability of opportunities to consult tutors	★★	★	OK	–	– –

25	The benefit you get from consulting tutors	★★	★	OK	−	− −
26	The fairness of the way your lab reports and assignments are marked	★★	★	OK	−	− −
27	The way in which your abilities have been assessed on this course	★★	★	OK	−	− −
28	The number of tests you have written up to now	★★	★	OK	−	− −
29	The extent to which the tests are a fair reflection of the standard of the work taught.	★★	★	OK	−	− −
30	The number of opportunities you have had to get feedback about your knowledge of this subject	★★	★	OK	−	− −
31	The behaviour of your fellow-students	★★	★	OK	−	− −
32	The level of knowledge of your fellow-students	★★	★	OK	−	− −
33	The opportunities you have had to work with and learn from other students	★★	★	OK	−	− −
34	The extent to which the instructor has inspired you	★★	★	OK	−	− −
35	Has your attitude towards the subject improved or worsened as a result of attending this course?	★★	★	OK	−	− −
36	How well you cope with the workload	★★	★	OK	−	− −
37	How has your experience of this subject influenced your attitude to engineering as a career?	★★	★	OK	−	− −
38	Do you feel you are using your time effectively in class?	★★	★	OK	−	− −
39	The design-and-build project you had to do	★★	★	OK	−	− −
40	The amount of support you think the department provided to the instructor	★★	★	OK	−	− −

Now, have a look at your responses to all the items in the above table, and if there are any issues that you would like to comment on, please describe them here:

What did you like best about this course?

What did you like least about this course?

Thank you for taking the time to give your instructor feedback about the way this course is meeting your needs. Please return this form promptly to your insructor, via your class rep.

Appendix 5

Other books by this author on basic engineering mechanics and on teaching

The series of three volumes: 'Basic Engineering Mechanics Explained' (2019)

These volumes set out in a reader-friendly way the essential principles of all the key concepts in basic mechanics and their application to a variety of situations in mechanical engineering.

This series would be found useful by any person either studying the subject, teaching it, or needing to apply it. That includes high school students intending to study engineering, first and second-year students at colleges and universities, and practising technicians and engineers.

The emphasis is on approaching problems with common sense, and reasoning from first principles. The level of mathematics used is appropriately unpretentious, because the author believes that it is more important for a student of mechanics to understand mechanical principles than to engage in high-level mathematics and programming.

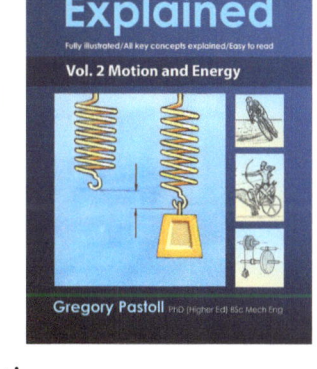

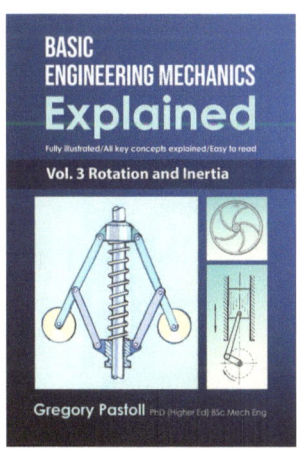

The basic principles of mechanics never change. If you understand them, you can grasp the essentials of any physical engineering situation.

The three volumes of Basic Engineering Mechanics Explained between them contain 1000 illustrations and over 420 original exercises and examples.

The solutions to many of the examples are shown with full workings. The answers to all questions requiring calculations are supplied.

Anyone who has mastered the principles that are set forth in that series will have an indisputably sound grasp of the science of mechanics, as applied to engineering.

Contents outline of this series

Volume 1: Principles and static forces

ISBN: Paperback 9780648466512; Hardcover 9780648466505; eBook 9780648466567

1 What Mechanics is about, and why we study it
2 Concepts, quantities, principles and laws
3 Working with numbers in engineering
4 Forces: components, resultants and equilibrium of a particle
5 Force moments, torque, equilibrium of rigid bodies, free-body diagrams
6 Centres of gravity and centroids
7 Forces in structures: frames and trusses
8 Friction between dry flat surfaces
9 Buoyancy

Volume 2: Motion and energy

ISBN: paperback 9780648466536; hardcover 9780648466529

10 Linear motion with uniform acceleration
11 Motion influenced by gravity: vertical and projectile motion
12 Rotary motion
13 Work, energy and power
14 Simple lifting machines
15 Inertia in linear accelerating systems
16 Linear momentum and impulse
17 Relative velocity

Volume 3: Rotation and inertia

ISBN 9780648466550; hardcover 9780648466543

18 Centrifugal and centripetal force

19 Rotational Inertia
20 Rotational and linear inertia in accelerating systems
21 Kinetic energy of rotation and angular momentum
22 Simple harmonic motion
23 Vehicle dynamics
24 Additional exercises, test questions and challenges

Revision Exercises in Basic Engineering Mechanics (2023)

ISBN for paperback: 9780645268881 and for eBook: 9780645268898

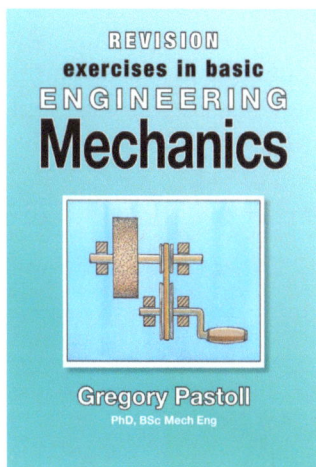

This book would be of use to students for self-study, and to instructors and examiners for the large variety of exercises it contains.

The exercises are all original, developed by the author in the course of his teaching, and reflect the practical applications of the subject in a variety of settings.

The topics in mechanics that are covered in these exercises comply with those listed in the above series. They cover the bulk of the topics found in most introductory courses in mechanics for engineers and technicians.

Included are:

- 45 ten-statement true/false quizzes, with answers, and a marking scheme that helps readers to identify exactly where more clarification might be needed, in order to consolidate their understanding of the principles of mechanics,

- 136 original, illustrated calculation exercises, with answers,

- 242 questions on all topics in mechanics, requiring short written answers, and

- 8 illustrated exercises suitable for tackling in groups.

All the exercises in this book may be used without seeking permissions.

Motivating People to Learn...and teachers to teach (2009)

ISBN 0-9585089-3-3 (The Attic Press) and 978-1-4389-1647-7 (Authorhouse), both paperbacks.

A comprehensive account of how to make use of intrinsic motivation to power the learning process.

Intrinsic motivation is that which comes from the experience of participating, as opposed to extrinsic motivation, which comes from the promise of success, in the form of prestige, jobs, money and privileges.

Using examples from many different educational settings, the author describes in detail twelve strategies that will help any teacher bring significant improvements to the learning experience of their students.

These twelve strategies are:

- Starting with the students' questions
- Making use of constructive competitiveness
- Giving students the right level of challenge
- Getting students to interpret original data
- Making the learning experience hands-on
- Allowing students the maximum freedom to play with ideas
- Shattering students' complacency
- Giving constructive feedback
- Providing students with the opportunity to excel
- Giving students the means to judge their own progress
- Co-operative learning techniques, and
- Exposing students to motivated people.

Appendix 6

Selecting the right mix of exercises to use for assessment

Having taught basic mechanics to first and second-semester engineering students for over 14 years, and having experimented with methods of assessment, I do not imagine that there is one 'best' way of assessing students' understanding of the principles of mechanics, that would satisfy everyone.

Instructors need to find the right mix for themselves and their students. In fact, I heartily support the right of instructors to teach and assess in whatever way they individually choose to do, as long as they can convince their HoDs that they are being professional and responsible, and are not compromising the standards of their institution.

If I were teaching mechanics to first and second-year students in a traditional 'teach, test, grade' system today, I would probably weight the components of the course grade approximately as follows:

Design-and-build project/s	20%
Lab test	15%
Short tests held after a written assignment or after tutorials	10%
Tests of prediction of mechanical consequences	5%
Average of two mid-term tests	20%
Final exam	30%

Each of the mid-term tests and the final exam might be designed so that the weighting of scores is in approximately these proportions:

Calculation questions	50%
True/false questionnaire, marked in the $2,1,0,-3$ method	20%
Short questions requiring written descriptive answers	10%
Definitions	10%
Questions requiring reasoning from first principles	10%

However, this arrangement represents only one person's take on how to assess students' knowledge of basic mechanics. Other instructors will do it differently.

To assess properly requires a significant and continuing input of energy from instructors. One has to keep on inventing new activities, new questions, new assignments, new projects and new criteria for assessment. There is no such thing as sitting back while assessment takes care of itself.

There is also no point in using the same assessment questions on every subsequent cohort, as students will naturally prepare themselves to be able to answer all recent examination questions.

So, assessing requires ongoing work from assessors. However, as with all activities in life, the more energy you put into it, the more likely you are to find satisfaction that it is working out the way you hoped.

What we don't test for, but possibly should

In the business of training people for the engineering professions, we usually focus on assessing their technical knowledge. Which is, of course, essential.

However, there are other characteristics that need to be displayed by people in these professions, besides knowing the science. We don't customarily assess these, but perhaps we ought. They are mentioned here as something to think about, because some of them are equally as important as technical knowledge in making a match between a person and this line of work.

In no particular order, these characteristics include:

- Exhibiting common sense (which is to some extent encouraged by some of the exercises proposed in this book),
- Ability to communicate technical ideas on the spur of the moment to other technical people verbally, as well as by writing and sketching (as opposed to learning formal communication protocols from non-technical communications specialists),
- Genuine interest in this field of endeavour (as opposed to taking a course for the sake of the prestige and monetary reward that is associated with being an engineer),
- Self-discipline (which is to some extent evident as a result of enduring *any* course of instruction and making the grade),
- Creativity in solving problems and in seeing what might be needed in the world, that engineers could contribute to creating,
- Ability to take the initiative in learning what he/she needs to know, and
- Reliability and level of responsibility shown when given tasks.

If all aspiring engineers were trained on the job, instead of in classrooms. candidates with these personal characteristics would soon rise to prominence.

However, the way modern society has developed works against finding apprentice-type placements for the number of people who currently attend colleges. So, it is left to educational institutions to filter out those candidates who are judged less likely to be successful, based on written evidence of *technical criteria* alone.

One past student of mine who was dyslexic, was unable to complete his diploma because he had difficulty in writing exams. Yet he was brilliant, and it showed all the way through his studies. He went on to become a very successful engineer, thanks to a company that recognised his abilities, without insisting upon a formal qualification.

We may not have a choice about whether or not to put candidates through courses of instruction. However, the least we can do is try to foster the type of thinking that is going to make students better prepared for the practicalities that they will face, and capable of using their initiative once they start work.

Appendix 7

Entrance testing vs exit testing

Collectively, across nations, many levels of education and almost all disciplines, the modern world has evolved a standardised pattern of 'teach, test, grade', where

- one or more people in a teaching institution teaches a course,
- the same people assess the students, then
- the institution puts its stamp of approval on 'successful' students so that they can go ahead supposedly 'prepared' for some further activity.

This further activity could be:

- attending a subsequent course in the same institution, or
- entering a higher institution, or
- taking up employment, or
- gaining membership of a professional body.

The pattern of 'teach, test, grade' is so ingrained that some of us would be hard pressed to find fault with it. And yet, this very pattern is responsible for many of the dissatisfactions that bedevil assessment.

Briefly, some of these dissatisfactions occur because:

1. If lecturers have to grade the students whom they themselves have taught, the results of that grading do not provide assurance that those lecturers have done a good job. They might have taught poorly, not completed the syllabus, or kept their expectations of the students low. They might have set and marked assessments in ways that give students an inflated grade, thus keeping up a misleading appearance of having been successful. When assessing student work, they may have put the emphasis on certain criteria that they feel are important, but that are not valued that highly by other participants in the process, and in particular, by those who will be employing their graduates.
2. There can't be an optimal relationship between the lecturers and their students. How can students fully trust a teacher who is going to have the final say over whether they are good enough? How can students be expected to provide honest feedback when responding to course evaluations, if the same lecturers who run those courses are later going to evaluate them? Also, it is difficult for a teacher to give fully of his or her knowledge and wisdom if teachers are predominantly preoccupied with assessment. Those students

who would have been interested in profiting from the rich resource of the teacher's experience don't customarily get sufficient time or opportunity to do so.
3. The stamp of approval provided by the institution always comes in the form of a grade, not a statement of what the student can actually *do*. So, those who teach a follow-on course or those who propose to employ the student don't really know what the student's capabilities are. They just have to believe that the institution that supplied the grade is reputable, and that the grade obtained actually *means* something. However, with each passing year, those grades are coming to mean less and less. An internet search for the phrase 'grade inflation' reveals widespread evidence of this phenomenon. At one institution where I taught, our department had had a good reputation for decades, but a colleague told me one day that one of his past students who was now an employer had phoned him and told him: 'We are no longer going to consider employing any of your fresh graduates. They are coming to us completely unprepared.' This was a sobering blow, but it had been a long time in coming, with many factors contributing to a drop in standards. So, what did our stamp of approval mean?
4. If generous allowances are continually made for the inadequate preparedness of some students, caused by factors like unrealistic high school grades, poor schooling, or social problems in their upbringing, this leads to lenient expectations and an inevitable drop in standards, with consequent grade inflation.

If the purpose of assessment is to testify that a student is ready for the next step, do the people whom the student will encounter in 'the next step' have any say about whether the students are properly prepared for what they will need to do when they get there? Usually not.

This is a real problem. How many times does it occur that students get through a course, only to be found wanting by those who have to teach the follow-on course? Haven't we all seen examples like the following?

- A student passes Physics 1, but can't explain to a lay person what 'relative density' means.

- A student enters college without knowing there is a difference between brass and bronze, or how to determine the density of a brick.

- A student graduates but shows no initiative when tackling practical problems in the workplace.

- Students knew what they were supposed to know, at the time that they were evaluated, but this knowledge has faded by the time they get to 'the next step'.

Their level of understanding was good enough to get through the exam, but didn't contribute to them building a knowledge structure that would last.

To counter all the negative aspects that arise from the standard pattern of 'teach, test, grade', I believe we need to completely flip our thinking around, because:

The ideal assessment situation would be one in which students have to do *entrance* testing rather than *exit* testing.

Passing an entrance exam definitely *does* ensure you are ready for the 'next step', because that 'next step' is going to occur right now, while the abilities revealed by the test are still present in the candidates.

To make a parallel with an engineering situation, suppose that your company relies on high-quality raw materials for its processes. If the suppliers are adamant that their products meet their specifications, but you regularly find that they don't meet yours, you have a problem. Wouldn't it be more reliable to dispense with your supplier's guarantees, and test the materials yourself before accepting them?

Applying this line of thinking to engineering instruction:

Suppose students want to enrol for my course in Mechanics 1. It would not matter to me whether they got an A or an E in high school, or whether or not they have studied physics, maths or materials. It would be of no consequence to me whether they came directly from school, or had lived and worked for some years since they finished school.

I would *like* them to have enough skills in the right areas, and in other areas in which have never been formally tested, but which are important, such as the ability to exhibit common sense. I'd like to know they have an aptitude for figuring out what is going on in a simple mechanics situation, reasoning from first principles, without applying formulas. However, I wouldn't know whether they had the requisite skills until I saw the way they handled their entrance test.

If they passed my entrance test, they would be ready to participate in my course.

Before taking the entrance test, any interested candidate would be given a list of what it would cover, showing examples of typical questions that could be asked. It would be up to them to prepare themselves for the test.

During their time with me, while studying Mechanics 1, they would have the opportunity to take the initiative, using my knowledge as a resource, to find out what they need to know so that they could take the entrance test for Mechanics 2. No matter what grade (as a feedback indicator) I gave them on my course, if they wanted to take Mechanics 2, they would still be required to take the entrance test for that subject. And so on, until, when they graduate, their prospective employers should make them take *their* entrance test.

There seems to be no set tradition for using the instructor's experience as a resource. What the instructor may or may not have done in an engineering career is usually not seen as relevant to a student's engagement with the course.

However, the following anecdote illustrates the potential value of making use of the instructor's experience, in a way that would not normally be thought appropriate on a conventional course:

In my fourth year of study, we had one particular lecturer who looked (and was) way beyond retirement age. He had suffered a stroke, and could not speak very well or loudly, as one side of his face was paralysed. He was from a country on the wrong side of the Iron Curtain, and his pronunciation was difficult to interpret. When he lectured, he clung to the rostrum, shaking and mumbling incoherently. We thought we were very badly done by, having this man to teach us.

So, we student representatives went to complain to the HoD, who listened to our side of the story, and then made us think. He explained that the man's family was still on the other side of the Iron Curtain (this was 1973) and depended entirely on the money that he was able to send them, so that compelling him to retire was not a moral thing to do. He pointed out that during WW2, this man had been commandeered to design locomotives for the Nazis. He had specialised in designing them so close to the limits of their performance that they would mysteriously fail soon after entering service. He also had 19 patents to his name. The HoD suggested that instead of expecting him to lecture, we might make use of the time in class to ask him questions about his career experiences relating to his teaching subject.

At our next class, we clustered close to the rostrum and started asking questions. The man came to life, and his subject came to life. We began to appreciate him and his knowledge, and to look forward to the fascinating stories from his experience. And we learnt a lot about the subject that we hadn't suspected existed.

An entrance testing system for employment would work if the selection of candidates was left to the engineering experts in the company, and not to the Human Resources Department, who can almost always be relied on to apply criteria that are politically motivated, and which have nothing whatsoever to do with competence.

An entrance-testing arrangement would be completely fair to all prospective entrants, if the criteria to be assessed in these entrance tests are spelled out publicly in advance, so that sufficiently motivated candidates can prepare themselves properly.

This author is well aware of the inertia that attends the way things are done in teaching institutions. I once heard a quotation to the effect that universities are capable of absorbing huge amounts of change, without changing in the least.

Therefore, I am under no illusion that the vast majority of teaching departments will stick to the traditional format of 'teach, test, grade'.

That is why the exercises that appear in my books are also able to be used for assessment by educators operating within an exit-testing system.

However, I urge readers to consider the advantages of moving away from the universally used exit-testing pattern, and adopting an entrance-testing system.

About the author

Gregory Pastoll taught basic mechanical engineering subjects to first- and second-semester students in a polytechnic/university of technology environment in Cape Town, specialising in the subject of Engineering Mechanics.

His interest in developing better teaching methods led him to take a position as a consultant on university teaching methods at the University of Cape Town.

He has a BSc Mech. Eng. from the University of the Witwatersrand, an M.Phil in evaluating student feedback about teaching, and a PhD in Higher Education from the University of Cape Town. His PhD research investigated the qualities that educators regarded as indicative of advanced educational development in a person.

While teaching, he experimented widely with ways of getting students to engage with the subject and to be motivated by interest.

He is the author of several books on mechanics and on teaching. See Appendix 5 for a full description of these books.

In addition to his career as an engineering lecturer, the author has also taught other subjects at different levels: English as a foreign language to adults in South Korea (one year) and in Austria (two years); mathematics and art to middle school children in Austria; and basic hand-tool woodwork and very basic mechanics to primary school children in South Africa.

He believes that teaching is mainly about motivating students, because motivated people will put their minds to anything they find interesting.

www.ingramcontent.com/pod-product-compliance
Lightning Source LLC
Chambersburg PA
CBHW041158290426

44109CB00002B/55